Challenges of the Agrarian Transition in Southeast Asia (ChATSEA)

Borneo Transformed

Agricultural Expansion on the Southeast Asian Frontier

T0345554

Challenges of the Agrarian Transition in Southeast Asia (ChATSEA)

The shift from rural societies dependent upon agricultural livelihoods to predominantly urbanized, industrialized and market-based societies is one of the most significant processes of social change in the modern world. The Challenges of the Agrarian Transition in Southeast Asia (ChATSEA) project examines how the transformation is affecting the societies and economies of Southeast Asia. Headed by Professor Rodolphe De Koninck, holder of the Canada Chair in Asian Research at the University of Montreal, and sponsored by the Social Sciences and Humanities Research Council of Canada, the ChATSEA project includes publications by senior academics as well as junior scholars.

Borneo Transformed

Agricultural Expansion on the Southeast Asian Frontier

Edited by

*Rodolphe De Koninck, Stéphane Bernard and
Jean-François Bissonnette*

NUS PRESS
SINGAPORE

© 2011   Rodolphe De Koninck, Stéphane Bernard and Jean-François Bissonnette

NUS Press
National University of Singapore
AS3-01-02, 3 Arts Link
Singapore 117569

Fax: (65) 6774-0652
E-mail: nusbooks@nus.edu.sg
Website: http://www.nus.edu.sg/nuspress

ISBN 978-9971-69-544-6 (Paper)

National Library Board, Singapore Cataloguing-in-Publication Data

Borneo transformed : agricultural expansion on the Southeast Asian
    frontier / edited by Rodolphe De Koninck, Stephane Bernard and Jean-
    Francois Bissonnette. - Singapore : NUS Press, c2011.
    p. cm.
    Includes bibliographical references and index.
    ISBN : 978-9971-69-544-6 (pbk.)

        1. Agricultural intensification - Borneo. 2. Agriculture - Economic aspects
    - Borneo. 3. Agriculture - Social aspects - Borneo. 4. Land reform - Borneo.
    I. Koninck, Rodolphe De. II. Bernard, Stephane, 1969- III. Bissonnette,
    Jean-Francois, 1983-

HD2075.8
338.1095983 -- dc22   OCN689939127

Typeset by: Forum, Kuala Lumpur, Malaysia
Printed by: Mainland Press Pte Ltd

Contents

List of Figures

List of Tables

Preface

This book is part of a series currently being prepared in the context of a five-year international research project (2005–10) funded by the Social Sciences and Humanities Research Council of Canada (SSHRC). Titled "Challenges of the Agrarian Transition in Southeast Asia", it involves collaboration between scholars, faculty as well as graduate students from Indonesia, Malaysia, the Philippines, Thailand, Vietnam, Singapore, Canada, the United Kingdom, France and Australia.

We therefore wish to thank not only the SSHRC, for its financial support for this project as well as an additional one devoted specifically to Borneo (2008–12), but also several individuals for their personal input. These include, first and foremost, Marc Girard, from the Geography Department at the University of Montreal. Marc is the creator of the majority of figures that illustrate this book. We also offer our thanks to the Cartographic Services at the Australian National University (ANU), headed by Kay Dancey. The initial versions of the figures that accompany Chapter 6 were drafted at ANU. At the University of Montreal, the initial versions of the six figures in Chapter 1 were drafted by Jean-François Rousseau. Monia Poirier, Bruno Thibert and Étienne Turgeon-Pelletier were also very helpful in the preparation of the manuscript and deserve our thanks. Finally, we thank Ramzah Dambul, Noboru Ishikawa, Fadzilah Majid-Cooke, Dimbab Ngidang, Jayl Langub, Tania Li, Patrick Sebat, Pujo Semedi, Jérôme Rousseau and Vincent Chai. Of course, none of these persons is responsible for any of the mistakes or inaccuracies that may be found in the book.

The editors

1
Southeast Asian Agricultural Expansion in Global Perspective

Rodolphe De Koninck
University of Montreal

Southeast Asia's Own Agrarian Transition

Since the 1960s, Southeast Asia's agricultural sector has been the object of phenomenal growth. Increases in agricultural productivity, in both food and industrial crops, have been and continue to be partly linked to the increasingly energy-intensive capitalization of agriculture and the rapid growth of agrifood systems and agribusiness, increasingly dependent on globalization forces.

This growth has also been largely associated with a combination or a convergence of two fundamental processes: agricultural intensification and territorial expansion, the latter appearing exceptionally dynamic in contemporary Southeast Asia. It is thus a key object of analysis in a major international study focusing on the specificities and implications of the region's agrarian transition. Titled "Challenges of the Agrarian Transition in Southeast Asia", the five-year research project (2005–10) aims to analyze and interpret a series of interrelated processes, including the two above-mentioned ones as well as market integration, industrialization and urbanization, population dynamics, intensification of regulation and, finally, environmental change (De Koninck 2004). All these processes are at play throughout the region, with particular intensity in Borneo. In addition, in the case of this large island located at the heart of Southeast Asia, agricultural expansion plays a central role, or so we hypothesize.

At first sight, the expansion of the region's agricultural domain may seem somewhat classical. Such a process has occurred throughout the world and continues today in several regions, as we will illustrate. But we also intend to show that, first, in Southeast Asia it plays an unusually and increasingly essential role in the dynamic transformation of the countryside and livelihoods of its inhabitants; and, second, the island of Borneo is currently at the core of this expansion process, with oil palm cultivation acting as the spearhead, following specific mechanisms and resulting in specific consequences that we intend to investigate.

A Brief History of Global Agricultural Expansion since the 18th Century

Throughout the world, until the 16th century cropland expansion remained moderate, as did population growth. Things began to change following European explorations of the 16th and 17th centuries, themselves followed by colonial expansion. By the early 18th century, cropland expansion along with population growth was already going strong, not only in Europe and Russia but also in the northeastern portion of the United States as well as in China (Ramankutty and Foley 1999) (Figures 1.1 and 1.2). Rapid agricultural expansion continued in all these regions during the first half

Figure 1.1 Estimated cropland in Europe and the Americas, 1700-1990

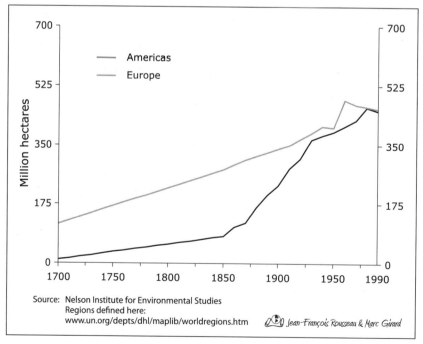

Source: Nelson Institute for Environmental Studies
Regions defined here:
www.un.org/depts/dhl/maplib/worldregions.htm   Jean-François Rousseau & Marc Girard

Figure 1.2 Estimated cropland in Asia, Africa and Australasia, 1700-1990

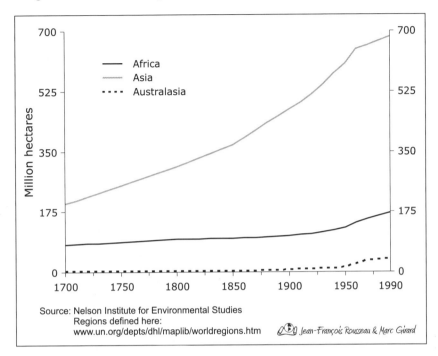

Source: Nelson Institute for Environmental Studies
Regions defined here:
www.un.org/depts/dhl/maplib/worldregions.htm   *Jean-François Rousseau & Marc Girard*

of the 19th century, but by the middle of the century it had begun to level off in Europe. The opposite occurred in the United States, where new land openings accelerated noticeably with the conquest of the West. In Russia, strong expansion lasted for centuries, also accelerating towards the middle of the 19th century, with even a massive spurt following a relapse during World War II. In China and in South Asia, while the 19th century witnessed a relative reduction in the pace of cropland expansion, this expansion remained strong until the middle of the 20th century. In China, expansion peaked at the time of the Communist takeover, but since the 1950s the country has been losing some of its agricultural land. The same has been occurring in Russia and Europe since the 1960s, but at a slower pace in the latter. As for South Asia, cropland expansion continued during the second half of the 20th century, with a noticeable reduction in pace after the 1960s.

Within South America, European-induced cropland expansion began at different times, depending on the region. On average, it remained very modest throughout the 18th and 19th centuries, accelerating noticeably only at the turn of the 20th century—first in the southernmost realm, particularly in the pampas, then in the northern countries, notably in Brazil, where expansion was massive from the 1950s until the 1980s. Cropland expansion has been somewhat comparable in Canada: modest during the 18th and 19th centuries, it accelerated during the 20th, but never at a pace comparable to that reached by countries to the south, whether the United States or Latin America. However, a reduction in agricultural land has been a characteristic common to most of the American realm, from Canada to the pampas, with the exception of Central America and, more recently, Brazil. In the United States, it actually began in the 1930s, with a few ephemeral expansion spurts since then. In the Central American domain, cropland expansion seems to have definitely levelled off in the 1990s.

In tropical Africa, the expansion of cropland remained modest throughout the 18th and 19th centuries and only began to accelerate noticeably by the 1940s. Since then, it has continued at a pace well in excess of the world average. The same applies, somewhat surprisingly, to North Africa and the Middle East, which have also been the object of sustained cropland expansion, particularly since the 1920s, with Turkey at the forefront.

With regard to mid- and late-20th century agricultural expansion, two other major regions stand out: so-called Pacific developed countries— essentially Australia and New Zealand—and Southeast Asia. In Australasia, agricultural expansion began noticeably during the last quarter of the 19th century. It accelerated slowly during the first half of the 20th, then very rapidly during the 1950s and 1960s. However, it has since levelled off.

As for Southeast Asia, its case seems rather unique: since the early 18th century, the region has been the object of uninterrupted agricultural expansion, the process having gone through two major bursts of acceleration. Fuelled by the expansion of Western imperialism, the first occurred at the end of the 19th century and continued for more than half a century. Along with post-World War II economic expansion, the second began in the 1950s and is still going strong, with apparently no equivalent in the world, except perhaps in Brazilian Amazonia.

Agricultural Expansion in Southeast Asia

During the 18th century, at a time when only the Philippines and limited portions of what was to become Indonesia were under colonial administration, cropland expansion in Southeast Asia remained relatively modest. According to Ramankutty and Foley (1999: 1019), between 1700 and 1800 cultivated land area grew by less than 30 per cent, from about 110,000 to about 140,000 square kilometres. During the first half of the 19th century, as population grew at an average annual rate of 0.7 per cent—from ~28 million to ~40 million (Neville 1979: 54)—cropland did not even keep pace, as it increased at an average annual rate of less than 0.3 per cent, covering a total of only about 170,000 square kilometres by 1850. During the second half of the century, the average annual rate of population growth accelerated substantially, reaching 1.5 per cent. Over the same period, the rate of cropland expansion also increased significantly but never surpassed the demographic growth rate. However, from the 1880s onwards agricultural expansion truly picked up. During the first half of the 20th century, as population continued to grow at an average rate of 1.5 per cent per annum, cropland expansion nearly matched that rate. Between 1950 and 1990, the rate of cropland expansion continued to accelerate, with the consequence that the area of cultivated land nearly doubled in the region.

However, over the same 40-year period the total population of Southeast Asia increased even faster, from 173 million to 437 million, multiplied by a factor of 2.5, with the average annual rate of demographic growth reaching some 1.7 per cent. Fortunately, overall food *production*, particularly rice, increased even faster, thanks to a substantial improvement in yields. This, of course, has been typical of agricultural growth in most of the world's regions. With the very rapid acceleration in the rate of population growth, the only way that food production has been able to keep up has been through significant increases in yield. This has particularly been the case in Southeast Asia (De Koninck and Déry 1997). However, along with massive increases in yield, Southeast Asian agriculture has also continued to expand territorially at a rate apparently unmatched anywhere else, except, once again, in Brazil (De Koninck 2003, 2006).

This said, from the late 19th century to the late 20th cropland expansion evolved in nature, going through those two exceptional phases of acceleration. The first began in continental Southeast Asia, with the last decades of the 19th century and the first decades of the 20th witnessing massive expansion of rice cultivation, largely in the deltas and lowlands of Cochinchina, Thailand and Burma. To this form of food crop expansion—predominantly fuelled by external demand—various types of commercial crop expansion, most noticeably rubber, were gradually added, particularly as the 20th century wore on. This occurred predominantly in Malaya and Indonesia. Consequently, the second acceleration in Southeast Asian cropland expansion, which began in the 1950s, was almost exclusively attributable to further investments in the cultivation of commercial crops, particularly rubber, coffee and oil palm, the last having definitely taken the lead since the 1990s. As a result, in comparison with the rest of the world, the Southeast Asian ratio of cropland expansion over population growth does appear unique (Figures 1.3 to 1.6).

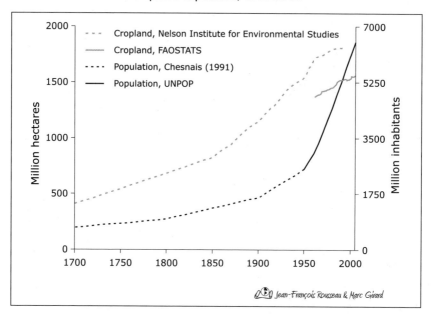

Figure 1.3 Estimated global population growth and cropland expansion, 1700-2005

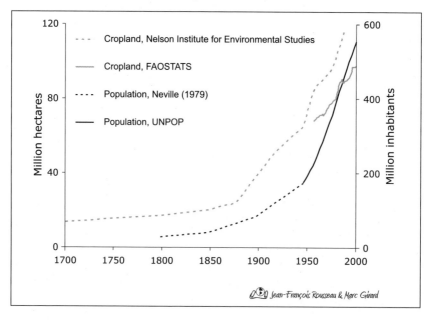

Figure 1.4 Southeast Asia. Estimated population growth and cropland expansion, 1700-2005

Figure 1.5 Southeast Asia. Cropland expansion by major region, 1700-2005

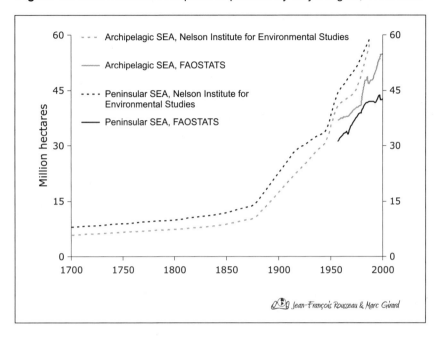

Figure 1.6 Southeast Asia. Cropland expansion by country, 1961-2005

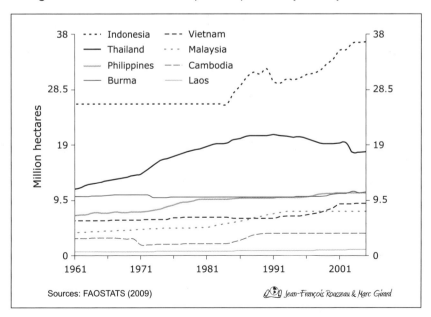

From the late 19th century onwards, Thailand's share in that expansion was dominant and remained so until the late 20th century. However, by the 1980s the rate of Thai agricultural expansion began to slow down, to such an extent that since the 1990s it has become negative, the only such case in the region (Figure 1.6). Cropland has continued to expand in all other countries in the region, with Vietnam, the Philippines, Malaysia and Indonesia taking the lead. Although the coffee boom in Vietnam has been particularly spectacular since the early 1990s, the current rate of growth does appear stronger in the archipelago, where the share of boom crops, and particularly oil palm, among all cultivated crops remains much higher. In that context, the case of Borneo Island, largely shared between Malaysia and Indonesia, appears exemplary. In fact, since the 1980s expansion within Malaysian territory has essentially occurred in the states of Sabah and Sarawak, while in Indonesia a large number of provinces have been involved, primarily on the islands of Sumatra and, increasingly, Borneo (Kalimantan). In nearly all cases, large private oil palm plantations have been or have become the main expansion agents. Consequently, while Borneo still has the largest remaining expanse of tropical rainforest in Southeast Asia, the area under cover is receding faster than anywhere else in the region (Langner *et al.* 2007). The island currently represents the region's largest and most active agricultural frontier. As such, it also represents a laboratory of agrarian transformations. As stated earlier, we intend to contribute to the analysis of some of the specificities and implications of these transformations. Ultimately, our approach is likely to be useful in the interpretation of transformations induced by the continuing expansion, well beyond Borneo, of boom crops, whether oil palm or others.

References

Chesnais, J.-C., *La population du monde de l'Antiquité à nos jours*. Paris: Bordas, 1991.

Dauvergne, P., "The Political Economy of Indonesia's 1997 Forest Fires", *Australian Journal of International Affairs* 52, no. 1 (1998): 13–7.

De Koninck, R., "Southeast Asian Agriculture since the Sixties: Economic and Territorial Expansion", in *Southeast Asia Transformed: A Geography of Change*, ed. Chia Lin Sien. Singapore: Institute of Southeast Asian Studies, 2003, pp. 191–230.

_____, "Challenges of the Agrarian Transition in Southeast Asia", *Labour, Capital and Society* 37 (2004): 285–8.

_____, "On the Geopolitics of Land Colonization: Order and Disorder on the Frontiers of Vietnam and Indonesia", *Moussons* 9–10 (2006): 33–59.

De Koninck, R. and S. Déry, "Agricultural Expansion as a Tool of Population Redistribution in Southeast Asia", *Journal of Southeast Asian Studies* 28, no. 1 (1997): 1–26.

FAOSTATS: http://faostat.fao.org/default.aspx.

GVM-TREES Project, *2000 Forest Cover Map of Insular Southeast Asia*. Ispra, Italy: Institute for Environment and Sustainability, Global Vegetation Monitoring Unit, 2002.

Langner, A., J. Miettinen and F. Siegert, "Land Cover Change 2002–2005 in Borneo and the Role of Fire Derived from MODIS Imagery", *Global Change Biology* 13 (2007): 2329–40.

Nelson Institute for Environmental Studies, University of Wisconsin-Madison, Center for Sustainability and the Global Environment-Sage: http://www.sage.wisc.edu/.

Neville, W., "Population", in *South-East Asia: A Systematic Geography*, ed. R. Hill. Kuala Lumpur: Oxford University Press, 1979, pp. 52–77.

Ramankutty, N. and J.A. Foley, "Estimating Historical Changes in Global Land Cover: Croplands from 1700 to 1992", *Global Biogeochemical Cycles* 13, no. 4 (1999): 997–1027.

Siegert, F. and A. Hoffmann, "The 1998 Forest Fires in East Kalimantan (Indonesia): A Quantitative Evaluation Using High Resolution, Multitemporal ERS-2 SAR Images and NOAA-AVHRR Hotspot Data", *Remote Sensing of Environment* 72, no. 1 (2000): 64–77.

UNPOP (United Nations Population Division), "World Population Prospects: The 2008 Revision Population Database", http://esa.un.org/unpp/index.asp.

White, M. and M. Klum, "Borneo's Moment of Truth", *National Geographic* 214, no. 5 (2008): 34–63.

World Atlas of Agriculture (1969–76). Novara: Istituto Geographico de Agostini, vol. 2.

http://maps.grida.no/go/graphic/extent-of-deforestation-in-borneo-1950-2005-and-projection-towards-2020.

http://www.panda.org/what_we_do/where_we_work/borneo_forests/publications/?21038/Map-Forest-cover-loss-in-Borneo-between-1900-and-2020.

2
Agricultural Expansion: Focusing on Borneo

Rodolphe De Koninck, Stéphane Bernard and
Jean-François Bissonnette
University of Montreal, University of Ottawa and
University of Toronto

"Each forest and village in Borneo had its distinct story of changes, but all also participated in some way in national and region-wide scenarios. The people and their environment on this island, long archetypes of isolation, exoticism, and singularity, shared much with their counterparts throughout Southeast Asia and the world beyond." (Padoch and Peluso 2003: 14)

Land, Forests and Peoples

The vast island of Borneo, in the heart of Southeast Asia, straddles the equator (Figure 2.1). Covering 743,000 square kilometres, it is larger than any country in the region, with the notable exception of Indonesia, to which it partly belongs. The granitic and mountainous core, which acts as a spine and structures the island, is oriented south-southwest/north-northeast. This is where the highest mountain ranges—Crocker, Iran, Moller and Schwaner—are found, with several peaks reaching over 2,000 metres in the Crocker Range and culminating with Mount Kinabalu at over 4,100 metres (Figure 2.2). The lower and less extensive Meratus Range is in the southeast portion of the island. The granitic core is largely absent from the southwestern portion of the island, which is characterized by swampy lowlands (Gupta 2005). In total, such lowlands occupy a much larger portion of the island than the mountainous areas. Uplands above 200 metres cover some 36 per cent of the total land area of Borneo, with 19 per cent above 500 metres. Corresponding figures for the whole of Southeast Asia stand at 54 per cent and 33 per cent. In short, the island of Borneo is significantly less mountainous than Southeast Asia as a whole and even than the rest of Indonesia and Peninsular Malaysia (De Koninck 2007).

As a result of its position on the path of both the southwest and wetter northeast monsoons, the entire island averages between two and three metres of rainfall per year, with a relatively large area in the mountainous core receiving more than four metres (Figure 2.3). In general terms, the central, northern and northwestern portions receive higher rainfall, Sarawak being particularly favoured, while the eastern portion of the island is susceptible to prolonged droughts (Dambul and Jones 2008). Rains are usually heavier during the northeast monsoon season, i.e., the last and first months of the year. Bintulu, located in the central lowlands, in a near coastal position, exemplifies this. Consequently, Borneo is endowed with a dense forest cover and an exceptionally intricate network of rivers, which practically all flow from its mountainous interior and radiate towards all shores (Figure 2.4). Unsurprisingly, an even denser hydrographical network covers the wetter north and northwestern portions of the island. Even so, the southern, Indonesian portion of the island is drained by very large rivers.

On both sides of the mountainous spine, many of these rivers are particularly powerful. This is the case with the Kinabatangan, Baram and Rejang on the Malaysian side and the Kapuas, Barito, Mahakam and Kayan on the Indonesian side. Along with their tributaries as well as numerous other smaller rivers, they represent key avenues of communication. In fact, in several regions waterways remain more important than roadways, servicing a much larger number of communities.

Borneo's forest heritage is exceptionally rich, biophysically as well as culturally. Cross sections from the coasts to the interior reveal a succession of forest formations, including mangroves in the intertidal zone, giving way to peat swamp forests and, farther in the interior, extensive areas of dipterocarp forests, of both the lowland and mountain types. Each of these is host to a variety of ecosystems rich in biodiversity, faunal as well as floral—in fact, Borneo's flora is "as rich as that recorded for the whole African continent" (MacKinnon and Sumardja 2003: 72)—with their current rapid demise being the object of major concern. More important, the big island's forests still are—or were until recently—inhabited by a large number of indigenous peoples, which makes for Borneo's distinctiveness. This is not to say that forest peoples do not exist elsewhere, but rather that on the island of Borneo they were and probably remain among the most diverse.

While the island's ethnic map appears quite complex and continues to be modified by migration flows, the general patterns identified by Sellato (1992) and Rousseau (1990) retain much validity today. The coasts and estuaries of major rivers are inhabited predominantly by Islamic Malay groups. Most of the Chinese are also found in these locations, particularly in the towns. In addition, Rousseau identifies "a lowland ring of shifting cultivators … [which] include the Ibanic people of the province of West Kalimantan and Sarawak, the Maloh of West Kalimantan, the Ngaju of Central Kalimantan and the Idahan speakers of Bulungan" on the eastern flank of the island (Rousseau 1990: 11). Rousseau distinguishes these lowland people from the inhabitants of the much more sparsely inhabited highlands of Central Borneo. Comprising agriculturalists as well as nomadic hunter-gatherers, the peoples of Central Borneo represent a dwindling minority of the island's population. Sellato has also provided a classification of the peoples of the entire Borneo Island (Sellato 1992) (Figure 2.5). It is largely compatible with the one suggested by Rousseau.

Territories, Populations and Settlements

The island of Borneo is shared among three sovereign states: Indonesia, Malaysia and Brunei (Figure 2.6). By far the smallest, the last is located on Borneo's northern shore, of which it claimed a much larger portion prior to

Figure 2.1 Borneo in Southeast Asia

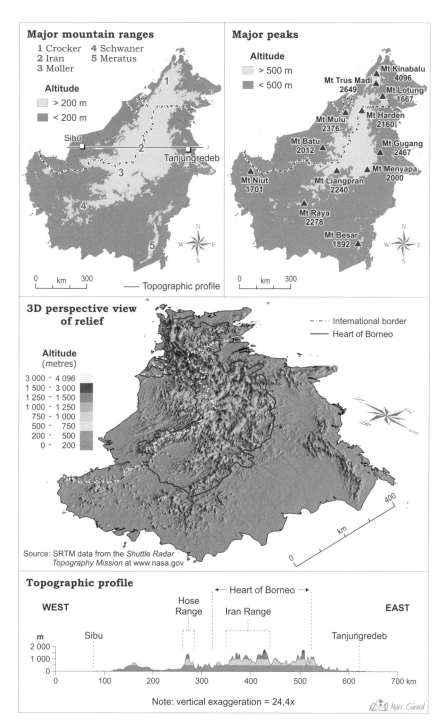

Figure 2.2 Uplands and lowlands

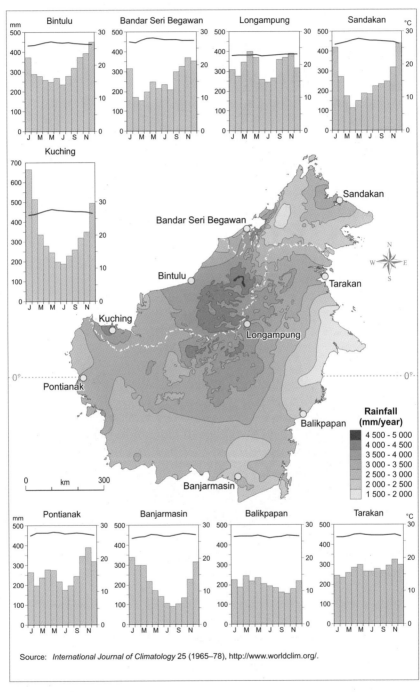

Source: *International Journal of Climatology* 25 (1965–78), http://www.worldclim.org/.

Figure 2.3 Rainfall distribution

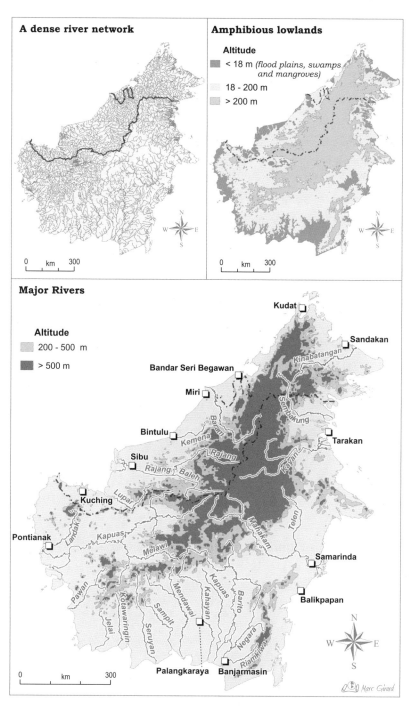

Figure 2.4 A high hydrographic density

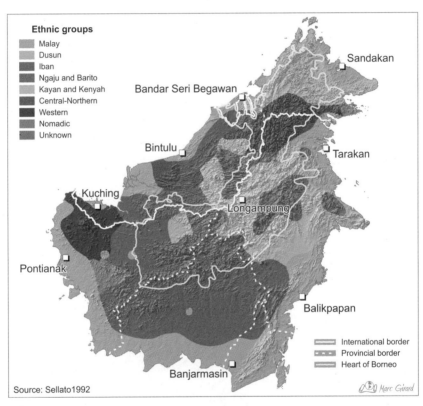

Figure 2.5 The peoples of Borneo

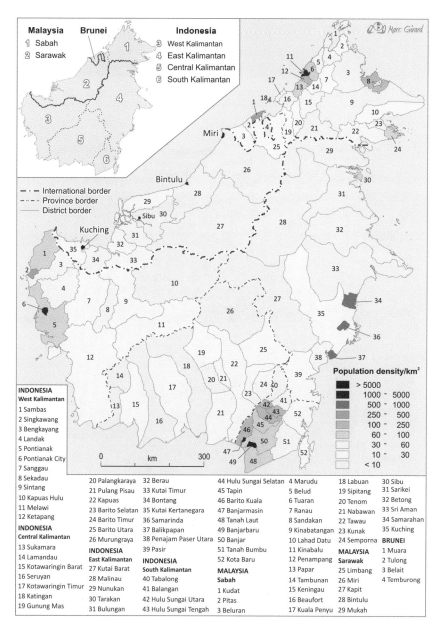

Figure 2.6 Population density by district and major cities, 2000

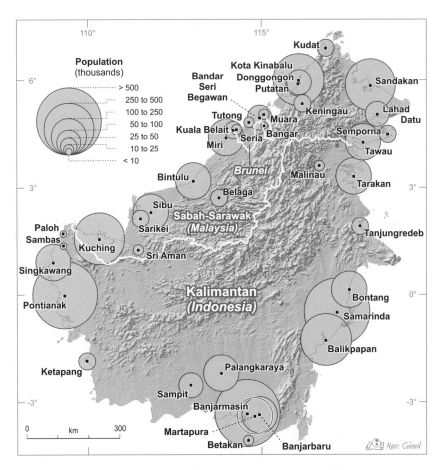

Figure 2.7 Cities and major towns, ~2000

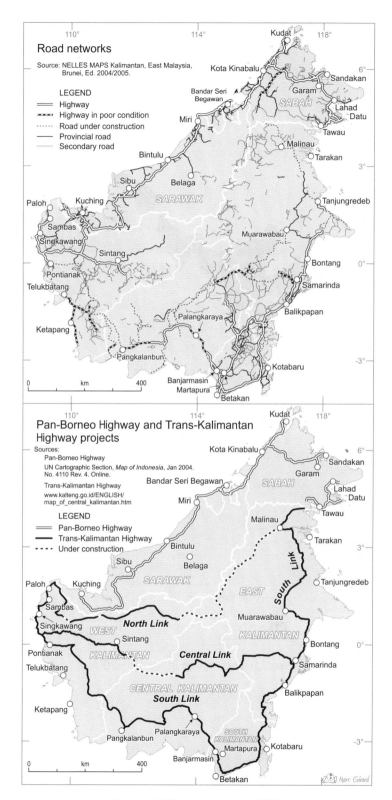

Figure 2.8 Road networks, ~2004

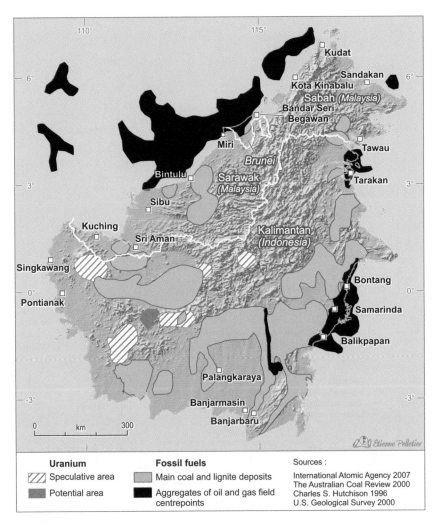

Figure 2.9 The mining industry, ~2000

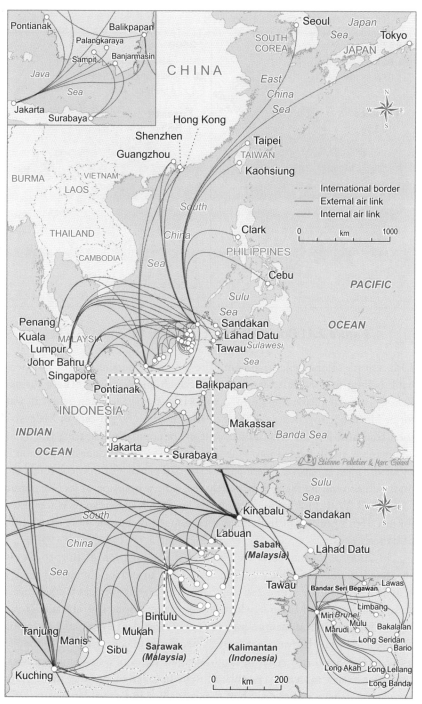

Figure 2.10 Internal and external air links, ~2006

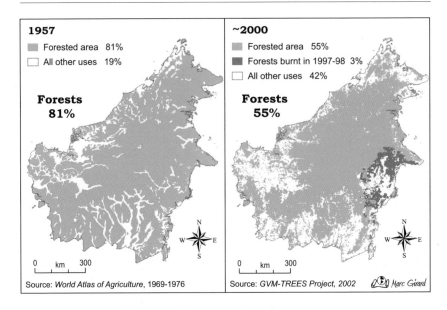

Figure 2.11 Borneo Island. Land use, 1957 and ~2000

19th century colonial encroachment. The remaining portions of the northern flank belong to the Malaysian states of Sabah and Sarawak. These former British colonial possessions joined the Malaysian Federation when it was formed in 1963. Together, in 2000 they accounted for 60 per cent of Malaysia's total land area but for only 20 per cent of its population. Bearing the name Kalimantan, itself divided into four large provinces, the Indonesian portion is by far the most extensive, covering more than 70 per cent of Borneo's total land mass and accounting for approximately the same proportion of its population (Table 2.1). With 11.3 million inhabitants in 2000, Kalimantan then accounted for just over 5 per cent of Indonesia's total population but for more than a fifth of its territory.

In 2000, according to census figures, the population of Borneo barely surpassed 16 million (Table 2.1). This resulted in an average population density of only 22 per square kilometre, well below the overall averages for Southeast Asia, Indonesia and Malaysia, which stood respectively at 124, 118 and 77 per square kilometre. Kalimantan and the much less densely populated Papua represented by far the two most sparsely populated major regions of Indonesia as well as of Southeast Asia. In the Malaysian states of Sabah and Sarawak, the latter even less densely populated than Indonesian Borneo, the average population density was one-sixth of that prevailing in Peninsular Malaysia (23 versus 134 per square kilometre).

Table 2.1 Borneo. Population, area and density, 2000

	Population (000)	Area (km²)	Density (inh./km²)
Sabah	2,468	73,620	33.5
Sarawak	2,010	124,449	16.2
Total Malaysian states	4,478	198,069	22.6
Total Brunei	333	5,770	57.7
West Kalimantan	4,034	149,700	26.9
Central Kalimantan	1,857	154,300	12.0
South Kalimantan	2,985	37,400	79.8
East Kalimantan	2,453	198,500	12.4
Total Indonesian provinces	11,331	539,900	21.0
TOTAL Borneo	16,142	743,739	21.7

Sources: Population and Housing Census of Malaysia 2000, Brunei Census 2001, Indonesian Census 2000.

However, within the various administrative divisions of the large island, population distribution remains highly diverse (Figure 2.6). The first, classical contrast lies between the coastal and interior regions. This applies throughout the island, notably in Sarawak, where the southwestern coastal lowlands show much higher densities than the interior. In Sabah, where the interior highlands have a long history of settlement (Lee 1965: 44), these appear today much less densely populated than the lowlands, particularly the narrow coastal plain that extends to the northeast of the Bay of Brunei. Even within the small Sultanate of Brunei, population distribution appears highly skewed. More than two-thirds of Brunei's inhabitants are found within the capital, Bandar Seri Begawan, and the surrounding coastal district of Brunei-Muara, which together account for less than a tenth of the sultanate's territory. In much larger Indonesian Borneo, the relative emptiness of the interior realm appears even more striking when compared with the clustering that occurs in the coastal lowlands, particularly in East Kalimantan as well as South Kalimantan. In coastal and estuarine districts, much higher population densities are more often than not attributable to urbanization, the larger urban establishments being usually located near the coasts or at river mouths, such as Bontang, Samarinda and Balikpapan in East Kalimantan (Figure 2.7).

Nevertheless, some relatively large urban settlements do exist in the interior, such as Palangkaraya, the capital city of Central Kalimantan, or even Sibu, a mid-estuary city on the Rejang River in Sarawak. Transport links between many of Borneo's urban settlements still rely primarily, if not entirely, on waterways. This is the case, for instance, with riverside settlements on the Kapuas and Mahakam Rivers, respectively in West Kalimantan and East Kalimantan, particularly in their upstream portions. In addition, in cases where road links do exist (Figure 2.8), these often remain functional only during the short dry season, with waterways still being much more reliable. In fact, several of the cities and towns situated in the swampy lowlands are accessible only by river or by air transport. This applies to Palangkaraya, on the Kahayan River. Although the city is located more than 100 kilometres inland, its population reached 160,000 in 2000. The growth of urban and rural settlements in Borneo is largely influenced by population migrations. While these have a long tradition on the island, contemporary in-migration from neighbouring islands is increasingly associated with Borneo's rapid growth as a resource frontier.

The Rapid Transformation of a Resource Island

In some ways, the landscapes and peoples of Borneo still have the distinctive appearance associated with the island's "exotic" image from past centuries. This can be attributed to the island's equatorial climate, along with its swampy and forested lowlands, often mountainous and even more extensively forested interior, rich fauna and flora, and, above all, its even richer ethnography. However, all of these have been changing rapidly—for decades, if not centuries (Padoch and Peluso 2003: 3). More important, since the 1960s, with their increasing integration into the national economies of Malaysia and Indonesia in particular, and into the world economy, the territories and peoples of Borneo have seen the pace of change quicken. Located at the heart of Southeast Asia, they have become increasingly involved in the key processes of change that have engulfed the region—they are in the eye of the storm, so to speak. These changes include the rapid growth of the mining and timber industries, followed by massive agricultural expansion, which in turn has contributed to an acceleration of in-migration, particularly from other Indonesian islands and the Philippines. Borneo has thus become the ultimate frontier territory within Southeast Asia. This also means that the island has been earmarked for mega projects in all of these sectors—mining, forestry and agriculture—the latter two often being related. Large hydroelectricity projects are part of this development. The Bakun Dam—the largest in Asia outside of China—is under construction in Sarawak, on the Balui River, a tributary of the Rejang. Although strongly contested, it is currently scheduled for completion in 2011. Plans are also advanced for the construction of several other dams on Sarawak rivers, including one 60 kilometres upstream from Bakun.

As for the mining sector, it has a rather long history in Borneo, with gold being mined by Chinese as early as the 1820s in western Sarawak (Jackson 1968: 144). Since then several other minerals, such as antimony, mercury and bauxite, have been extracted, but none has been as important as fossil fuels and uranium. Nowadays, these represent major resources for Brunei as well as Sarawak, Sabah and Kalimantan (Figure 2.9). The last is rich in coal and uranium as well as oil and natural gas, while the northern flank of Borneo, including offshore, is particularly well endowed with the latter two. In fact, the wealthy Sultanate of Brunei relies almost exclusively on its petroleum and natural gas resources, while Sabah and Sarawak are essential contributors to Malaysia's own prosperous petroleum and natural gas industry. In Sabah and Sarawak on the one hand, and East Kalimantan on the other, the considerable returns on petroleum and natural gas are the subject of much negotiation between the local states and the central ones, primarily represented by their respective national companies, Petronas and Pertamina. In all cases the industry has an essentially enclave character.

The forest resources of Borneo have been even more significant. Their collection for export, particularly products such as camphor and other forest produce, goes back several centuries (Cleary and Lian 1991). In fact, "the commercial exploitation of forest products to satisfy international tastes is at least two thousand years old" (Padoch and Peluso 2003: 4). After World War II, this exploitation began to accelerate significantly. By the late 1940s, external demand for Borneo timber helped the timber industry to take off. Sabah was to take the lead, particularly after large forest concessions were granted to timber companies in the 1950s, and even more since the 1960s, when production expanded rapidly and contributed to equally rapid deforestation.

By then the function of the entire island as a resource frontier for both Malaysia and Indonesia appeared even more evident. As for Sabah, it had become the main saw log exporter in the entire Indonesia-Malaysia region (Brookfield and Byron 1990). By 1970, the forestry sector represented close to 70 per cent of state revenue. Then, Sarawak gradually took over as the leading Malaysian state for the production and export of saw logs, the forestry sector providing more than half of state revenue in 1995. On the Kalimantan side, the logging boom had begun in the late 1960s, with the result that "The island of Borneo alone has provided at least one half of the world's total exports of non-coniferous tropical hardwoods since the early 1970s and 60 per cent in 1987" (Potter *et al.* 1995, referring to Brookfield and Byron 1990). In other words, as in the case of the fossil fuel industry, Borneo has been contributing more than its share to the national coffers of Malaysia and Indonesia, as well as the coffers of local and national timber barons. More recently, timber extraction has proceeded apace, but with an increasing proportion of the wood being processed in local mills, for example in Kalimantan, due to the ban on log exports in support of the plywood industry.

The timber industry has also found a new ally, namely, the oil palm industry. In both Malaysian and Indonesian Borneo, particularly since the 1970s, the spectacular expansion of oil palm cultivation has created a demand for ever more cleared land, which the foresters have been happy to provide, to the point of clearing much more land than needed, particularly in Kalimantan. It is the analysis of this very process of agricultural expansion and of its role in the overall agrarian transition that lies at the core of this book.

The road as well as air transport networks eloquently illustrate the partitioning of the island into separate political entities (Figures 2.8 and 2.10). Besides the road linking Kuching in Sarawak with Pontianak in West Kalimantan,[1] there are very few road links between the Malaysian and Indonesian sides and there is still no proper pan-Kalimantan road network. (The only railroad on the island consists of a 134-kilometre-long single-track line, in Sabah; trains, which cover the distance in some seven hours, cater primarily to tourists.) However, on both sides of the border, particularly on the Malaysian one, roads are rapidly being improved. Along the northern shores of the island, an all-season, well-surfaced highway links the major cities within each of the two Malaysian states. However, Brunei lies between them. Even though the sultanate itself is equipped with good roads, for political reasons Malaysian traffic cannot flow easily through it. Nevertheless, the Malaysian Pan-Borneo Highway is much closer to reality than the Trans-Kalimantan Highway. On the Indonesian side of the border, plans to improve and link up a number of roads, some of which are still in dire condition, have some way to go (Figure 2.8).

While there are a large number of airports on the island, most of these are for the use of small private companies. As for scheduled airlines, their primary function consists more in linking the island's respective components with the exterior than among themselves. For example, there are virtually no official air links between the Malaysian and Indonesian sides. However, regular flights link Kota Kinabalu and Kuching, the capital cities of Sabah and Sarawak, along with an increasing number of smaller cities within the two Malaysian states. There are no such air links between the capital cities of the Indonesian provinces—namely, Pontianak, Palangkaraya, Banjarmasin and Samarinda—and the command city of the petroleum and gas industry, Balikpapan. On the other hand, all of these cities, particularly Pontianak and Balikpapan, are serviced by regular flights to Jakarta. Similarly, Kuching and Kota Kinabalu airports are serviced by several daily flights to Kuala Lumpur as well as to Singapore. In addition, largely because of the rapid growth of the tourism industry in Sabah, Kota Kinabalu now maintains direct air links with Hong Kong, Taipei, Seoul and Tokyo. In short, while the national and international integration processes of large sections of Borneo are intensifying rapidly, the island's internal linkages seem to be lagging behind, particularly in Kalimantan, to a degree rarely seen throughout Southeast Asia (Rimmer 2003).

As these external linkages have expanded, so has the island's population. In recent decades, Borneo's role as a resource frontier has been increasing, and so has the number of migrants from neighbouring islands of Indonesia and the Philippines, with labouring on plantations and various construction sites being the dominant attraction. Consequently, the population of Borneo has grown more rapidly than that of the rest of Indonesia or of Malaysia, and it continues to do so. According to official figures, between 1961 and 2000 Borneo's population tripled, from 5.4 million to 16.1 million inhabitants (Table 2.2), while the population of the whole of Indonesia was multiplied by 2.1 (from 97 million to 206 million). During the same period, Sarawak's population increased by a factor of 2.7 and Sabah's by a factor of 5.4. In reality, the latter has grown even faster than these official figures reveal, as a substantial proportion of migrants enter Sabah illegally, many of them from the Philippines. Finally, within the big island, particularly between its Indonesian

Table 2.2 Borneo. Population, 1960–2000

	1960[a]	*1970*[b]	*1980*	*1990*[c]	*2000*
			(000)		
Sabah	454	653	956	1,737	2,468
Sarawak	745	976	1,236	1,648	2,010
Total Malaysian states	1,199	1,629	2,192	3,385	4,478
Total Brunei	84	185	200	260	333
West Kalimantan	...	2,020	2,486	3,229	4,034
Central Kalimantan	...	702	954	1,397	1,857
South Kalimantan	...	1,699	2,065	2,598	2,985
East Kalimantan	...	734	1,218	1,877	2,453
Total Indonesian provinces	4,102	5,155	6,723	9,101	11,331
TOTAL Borneo	5,385	6,969	9,115	12,746	16,142

Notes: a: 1961 in Indonesia; b: 1971 in Indonesia; c: 1991 in Malaysia and Brunei.

Sources: Population and Housing Census of Malaysia, various years; FAO Stats, Indonesian Census, various years; Jones 1964.

Table 2.3 Sabah, Sarawak and Brunei. Major ethnic groups by
percentage, 2000

Sabah	
Malay	15.0
Chinese	13.3
Kadazan-Dusun	18.0
Bajau	17.0
Other (~25) groups	36.7
TOTAL	100.0
Sarawak	
Malay	23.8
Chinese	26.2
Iban	30.1
Bidayuh	8.2
Melanau	5.7
Orang Ulu[a]	5.8
Other groups	0.4
TOTAL	100.0
Brunei	
Malay	66.3
Chinese	11.2
Indigenous[b]	3.4
Other groups	19.1
TOTAL	100.0

Notes: a: Includes, among others, the Kenyah, Kayan, Kedayan, Kelabit, Bisaya
and Punan; b: Includes, among others, the Kedasan, Murut, Dusut and
Iban.

Sources: Population and Housing Census of Malaysia 2000, Brunei Census
2001.

and Malaysian components, labour migration is also prominent. Consequently,
when actual numbers and percentages are taken into account (Table 2.3),[2] the
complex ethnic composition of the island's population appears quite different
from the one illustrated earlier (Figure 2.5).

For example, on the Malaysian side the relative numerical importance of
the Malays and Chinese appears noticeably less significant in Sabah than in
Sarawak, where the Iban are more numerous than either of these two major
groups. In Sabah, in fact, the population's ethnic composition appears more
complex, even if official figures do not reveal the importance of Indonesian
and Filipino migrants. However, in the case of Kalimantan the official census

figures (Suryadinata *et al.* 2003)—even if contested by many (Achwan *et al.* 2005)—do reveal the importance of migrants from other islands, thereby underlining the agricultural frontier character of much of Borneo. These figures indicate, for example, that in each of the four Kalimantan provinces, migrants from Java (whether Sundanese or, predominantly, Javanese) account for at least 10 per cent of the population. These migrants account for as much as 31 per cent in East Kalimantan (Table 6.2).

Between 1960 and 2000, according to census figures, the total population of Borneo Island tripled, from nearly 5.4 million to nearly 16.1 million inhabitants. Over the same period, according to estimates, total cultivated land slightly more than tripled, from 1.8 million to 5.6 million hectares. Both growth ratios appear slightly higher than the equivalent ratios for Southeast Asia as a whole, where, between 1950 and 1990, population and cropland grew by a factor of 2.5. Borneo's higher than average demographic growth was attributable to in-migration, which on the Kalimantan side was largely attributable to the Indonesian transmigration programme. On both the Indonesian and Malaysian sides, these migrants, including spontaneous ones—particularly in Sabah—were predominantly attracted by the growth of the mining, timber and agricultural sectors and, ultimately, urban expansion itself.

In the two Malaysian states and in all four Kalimantan provinces, a dominant proportion of agriculture's territorial expansion has been achieved through oil palm planting. In Sabah, the spread of oil palm cultivation took an early lead over the other regions of the island (Table 2.4) beginning in the late 1970s. However, by the late 1990s both Sarawak and Kalimantan, particularly the latter's western and central portions, had joined in the rapid expansion. While in 1990 oil palm cultivation covered some 400,000 hectares of Borneo's territory, by 2007 it had surpassed 3.6 million hectares. In addition, plans for further expansion appeared very ambitious throughout the entire island.

Expansion of cropland does not mean only that forests are replaced by agricultural fields or that some crops give way to others. It also implies more complex transformations of land use, as when forests are cut down, supposedly to give way to agriculture, but instead are simply left fallow. This is a process very different from the one involved in traditional swidden agriculture. Yet it represents a current occurrence in Borneo, particularly on the Indonesian side, where, after forests are intentionally or accidentally burned down—sometimes over vast expanses of land—no form of agriculture takes over. The comparison of land-use maps of the whole island for 1957 and ~2000 (Figure 2.11) clearly illustrates this. Although a number of land-use maps of Borneo Island have been published over the years, few are fully reliable and, equally important, they remain difficult to compare. Nevertheless, the World Wild Fund for Nature

Table 2.4 Borneo. Oil palm-planted area, 1975–2007 (ha)

	1975	1980	1985	1990	1995	2000	2003	2005	2007
Sabah	59,139	93,967	161,500	276,171	518,133	1,000,777	1,135,100	1,209,368	1,278,244
Sarawak	14,091	22,749	28,500	54,795	118,783	330,387	464,774	543,398	664,612
Total Sabah/Sarawak	73,230	116,716	190,000	330,966	636,916	1,331,164	1,599,874	1,752,766	1,942,856
Total Kalimantan	0	1,450	9,514	87,092	280,247	809,020	1,006,878	1,161,599	1,661,007
Total Borneo	73,230	118,166	199,014	418,058	917,163	2,140,184	2,606,752	3,379,577	3,603,863
Total Malaysia	641,791	1,023,306	1,482,399	2,029,464	2,540,087	3,313,393	3,802,040	4,051,374	4,304,913
Total Indonesia	188,825	294,560	597,362	1,130,000	2,020,000	4,158,700	5,250,000	5,453,817	6,783,000
Sabah/Malaysia %	9.2	9.2	10.9	13.6	20.3	30.2	29.9	29.9	29.7
Sarawak/Malaysia %	2.2	2.2	1.9	2.7	4.7	10.0	12.2	13.4	15.4
Kalimantan/Indonesia %	0.0	0.5	0.6	8.0	13.8	19.5	19.2	21.3	24.5

Sources: Malaysian Palm Oil Board (MPOB) 2008; *Perkebunan Dalam Angka 2008*; *Kalimantan Barat, Tengah, Timur, Selatan Dalam Angka*, relevant years; BPS Jakarta, July 2008; Departemen Pertanian, Direktorat Jenderal Perkebunan, Jakarta 2007.

(WWF), through its campaigns to alert the world to the plight of the Borneo rainforest and its wildlife, particularly the orang utans, has produced several series of maps illustrating the retreat of the forest. One of the more commonly used of these series consists of a set of six bicolour maps representing the extent of the big island's forest cover in 1950, 1985, 2000, 2005, 2010 and 2020. Another WWF map, more recently published, illustrates forest cover loss between 1900 and 2000, with a projection to 2020. In both series, the representations of forest cover for 2000 and 2020 are identical. Unfortunately, WWF does not indicate the sources utilized for the construction of these maps, nor does it provide a working definition of forest cover. In addition, a number of contradictions appear, particularly between the 1985 and 2000 maps, the ones most likely to have been based on satellite imaging interpretation—not an easy task when a cloud-covered territory such as Borneo Island is concerned. As for the 2005 map, it is probably a projection, just as, of course, the ones for 2010 and 2020. Notwithstanding these caveats, the WWF maps deliver a clear message: the forests of Borneo are receding at an accelerated pace.

This is corroborated by the comparison of two well-known land-use maps of the island, based respectively on 1957 and ~2000 data. The first is drawn from the *World Atlas of Agriculture*, the plates of which were published from 1969 through to 1976; and the second from the *2000 Forest Cover Map of Insular Southeast Asia*, published in 2002 by the GVM-TREES project. However, although these maps do distinguish between various forms of land use, they categorize and represent them quite differently, which renders a detailed comparison nearly impossible. Furthermore, the mapping techniques used for the 1957 map, based on air photo interpretation, were much sketchier than those employed for the construction of the *2000 Forest Cover Map*, derived from satellite imagery, more precisely SPOT 4 Vegetation imagery.[3] For those reasons, the representation utilized here (Figure 2.11) is simply bicolour, just like the maps published by the WWF, it being understood that the sources used are clearly identified and reliable. In fact, the *2000 Forest Cover Map of Insular Southeast Asia* (GVM-TREES Project 2002) has become a standard source (Achard *et al.* 2002) for all scholars interested in the fate of Borneo's land use and, more fundamentally, its inhabitants.[4]

All this said, the comparison of the extent of forested land in 1957 and ~2000 confirms what the WWF map series has illustrated and several authors have been warning against (Potter 2003, MacKinnon and Sumardja 2003). In just over 40 years, Borneo's forests have receded significantly, their extent declining from 81 per cent to 55 per cent of the total surface of the island.[5] The retreat has occurred mostly in the lowlands, particularly in Sabah and even more in Kalimantan, where deforestation has been extensive in all provinces.

The large "blank" that appears in East Kalimantan on the ~2000 map illustrates the devastation inflicted by the fierce forest fires that raged throughout the region in 1997 and 1998, the total area thus "cleared" of its forest cover amounting to just over 3 per cent of the total surface area of the island. Largely attributable to a combination of logging, plantation clearing and El Nino years, these fires—neither the first nor the last to have ravaged large portions of the island—did generate a host of criticisms as well as interpretations of their environmental and economic consequences (Dauvergne 1997, Siegert and Hoffmann 2000).

Since then, the retreat of the forest has continued, to a point where the mountainous core of the island seems to be under siege (Figure 2.11). The deforestation has gone hand in hand with agricultural expansion, mainly of oil palm, the boom crop *par excellence*. Thus, Borneo truly stands in the eye of the storm.[6] It is the implications of this storm of agricultural expansion that we wish to explore, but not before providing a brief overview of the literature on contemporary Borneo.

Current Issues in Assessing the Agrarian Transformation of Borneo

In Malaysian and Indonesian national initiatives Borneo is often portrayed as a periphery, a marginal domain, taking the form of a resource frontier. While that conception remains valid, the island should also be considered as the centre of major transformation processes associated with agricultural expansion, particularly oil palm expansion, occurring throughout much of Southeast Asia and possibly other regions of the world.

In recent years, several authors have offered insightful accounts on aspects of transition and environmental change in Borneo, including Padoch and Peluso (2003), Majid-Cooke (2006), and Sercombe and Sellato (2007). We wish to draw attention more specifically to spaces created by contemporary agricultural expansion from the 1960s onward. As pointed out in Chapter 1, the concept of agricultural expansion forms the basis of our analysis, as it refers to a key process around which fundamental issues of agrarian change revolve in contemporary Borneo. Of course, as also stated earlier, trade in forest and mineral products along with socio-environmental change is not new to the island. However, in the recent past, not only has agricultural expansion occurred at an unprecedented pace on the whole island, but it has also involved new actors, broadening, even globalizing, the issues. Spearheaded by the logging industry, booming cash-crop economies have remodelled territorial dynamics. Yet, we do not imply that agricultural expansion has been a uniform process, has occurred independently of other processes, or has served a single purpose.

Our goal is rather to show the complexity and diversity of local and state forms of agricultural expansion and the specific nature of the transition that they spearhead along with the problems that they generate.

Studies on agrarian and environmental change in Borneo have generally been on a regional level, for example, Sarawak, Sabah or a province of Kalimantan. Often related to the broad field of political ecology—although not always designated explicitly as such—these have focused mainly on institutional actors and their impact upon rural populations, providing additional insights on struggles over resources. Contributions by political scientists, social and cultural anthropologists, and human geographers have also generally had a local focus (Dandot 1992, Casson 2006). Largely based on fieldwork and the study of institutional settings, these important contributions have generally taken the form of case studies (Dove 1986, Ngidang 2002, Eghenter 2005). Such localized endeavours have enriched our understanding of local processes of agrarian change and demonstrated the impact of state regulations on the livelihoods of peasant groups. Recently this analytical framework has been expanded with the identification of the local impacts of global trends (Potter 2005, Bernard 2006). Breaking away from more traditional studies bounded by state entities, a growing body of research is also emphasizing cross-border ethnic links and migrations (Ardhana *et al.* 2004, Ishikawa *et al.* 2005, Wadley and Mertz 2005). This literature points to the importance of locating contemporary transformations in Borneo in a broader historical and geographical context. Many studies have contributed to highlight the issues at stake in current social and environmental transformations in Borneo. We provide here a brief overview of these issues.

With reference to recent scholarly work on Borneo, particularly since the early 1980s, four main categories of empirical research appear of particular relevance to our contribution on agricultural expansion. These are not independent of one another, nor are they bound to Borneo alone. Rather, they often concern the global South.

(1) By the beginning of the 1990s, logging, deforestation and ensuing land-use changes were accelerating in most regions of the developing world, notably in Borneo. This situation spurred research initiatives to understand the role of the state as well as local dynamics in these processes. As forest resources dwindled and crops such as oil palm expanded, the issue of land access became the object of renewed research (King 1993; Brookfield, Byron and Potter 1995; Majid-Cooke 1996; Drummond 1997; McCarthy 2000; McMorrow and Talip 2001; Jomo *et al.* 2004; Potter 2005). Deforestation issues per se, along with the rising global awareness of environmental transformation, gave birth to investigations on conservation initiatives and

agroforestry systems. These initiatives induced researchers to look into local actors' involvement in environmental change and to raise environmental sustainability issues.

(2) With respect to the highly political nature of agricultural expansion, which necessarily implies transformations in land access, modes of property and often enclosure, land rights are highly relevant to current agricultural expansion in Borneo. Several studies have addressed more specifically historical and contemporary legal aspects of the land rights of rural populations and the scope of claims on customary territories. They have highlighted several ethical aspects of land regulations, access rights and the arbitrariness of state policies (Colchester 1993, Colchester et al. 2006). They have also dealt with legal definitions of communities and customary lands (Cleary and Eaton 1992; Durand 1999; Doolittle 2001, 2003). Institutional shifts and legal definitions of the entitlements of indigenous populations to natural resources have been investigated, often in an attempt to expose hidden interests and agendas (Appell 1995, Ngidang 2005).

(3) A focus on the political nature of agricultural expansion in Borneo has highlighted the significance of the role of policy initiatives and agrarian change induced by government interventions in Borneo, whether through economic development or environmental protection. The agency of foreign actors, institutions and NGOs has also been the object of greater attention (Majid-Cooke 2002, 2003; Rosser et al. 2005). Since the fall of President Suharto in 1998, fundamental changes at the political level—notably, decentralization—have occurred throughout Indonesia, and these have greatly influenced resource allocation, especially in Kalimantan (Rhee 2003, Levang et al. 2007). This has in turn engendered new fields of investigation on natural resource access and entitlements (McCarthy 2004, Warren 2005). However, while migration, resettlement and regulations controlling people's movement have also been determinants of socio-environmental transformation, particularly through agricultural expansion, they have yet to be studied with sufficient attention.

(4) Understanding how local people manage their environment in different contexts, how they shape and reshape their interactions with the natural environment, and how they manage the local economy has remained essential to the appraisal of socio-economic and environmental change in Borneo (Hong 1989, Brosius 1997). The emergence of new technologies and crops has brought about large-scale environmental transformation. Resource exchanges and utilization, off-farm work and income diversification are encompassed in these studies (Dove 1986, 1993; Cramb 1985, 1993; Morrison 1993; Wadley and Mertz 2005). The controversy surrounding

the environmental impacts and economic significance of shifting cultivation in Borneo has consistently come to the fore, raising fundamental questions about environmental systems (Freeman 1970, Padoch 1985, Cramb 2007). Local-scale land use as well as global-scale processes are very important in the analysis of overall agricultural expansion in contemporary Borneo.

Our Perspective, Objectives and Contributions

Taking into consideration previous work, as well as the current dynamics of environmental transformation in Borneo, we contend that agricultural expansion impacts not only on territories but even more on the societies and livelihoods it supports. We intend to take into account the multiple livelihood and environmental changes, largely induced by state and global capitalism, in what is still a predominantly agrarian setting. This implies an examination of the links between dynamics at different scales and time periods. In this regard, the processes under scrutiny take place in a space where colonial and postcolonial regimes have reshaped long-established power relations. In addition, we consider the biophysical environment not only as an arena in which negotiations and struggles over resources take place, but also as an intrinsic component of the stakes. In fact, we view environment as an essential element in socio-economic relationships.

To cope with the dynamics of social, economic and political change, local populations have transformed land and forest management practices, as well as the meaning of nature itself. Land, which used to lie at the core of rural dwellers' livelihood, is now increasingly conceptualized as one economic component among others. Nevertheless, agriculture obviously remains a fundamental component of Borneo's economic system. As territories are the substrate of societies and the locus of state power, expansion of agriculture has major socio-economic implications. In this regard, it involves countless actors, themselves driven by a variety of motives. Therefore, our endeavour establishes links between different geographical scales of socio-environmental change, local, national and global, in order to better understand factors fuelling the territorial transformation of Borneo. According to Gupta (1998), global forces are not only located beyond the frontiers of the state, but also embedded within state and local processes. Global demand for agricultural goods affects rural communities and can influence or spearhead state strategies of national integration.

Populations of Borneo have not evolved in a pristine space but rather have been progressively integrated within state structures. Regulations elaborated by states to control territories, resources and populations have, in fact, moulded relations between people and the environment, drawing borders and defining access to and exclusion from territories. This process was identified by De Koninck (1981a, 1981b) and by Vandergeest and Peluso (1995) as state territorialization. The latter comes into conflict with the very dynamics of local or customary territoriality, which in effect it attempts to negate or at least to supersede (Raffestin and Bresso 1976; De Koninck 1981b, 1983).

State initiatives, policies and coercive power still have far-reaching implications within the political entities of Borneo, but they take different paths depending on the state involved. Moreover, most forms of development currently promoted by the state entail processes associated with globalization and the expansion of neo-liberal ideologies and policies. In addition, increasing market integration of communities located at the periphery of the world economy, far from urban centres, involves other forms of regulation. These regulations are as significant as the territorialization power of the state but derive mainly from the demands of the population. While market integration has often been facilitated by state agricultural development schemes in Borneo, local communities themselves can also deliberately choose it, as the economic importance of commercial crops is progressively appreciated and readily integrated into local agrosystems. In addition, with deregulation increasingly related to when not dictated by globalization, the level of global integration of commodity markets appears unprecedented. In fact, processes of the commoditization of vital aspects of human life—land and labour—and their corollary, enclosure, privatization and sometimes dispossession, reveal the far-reaching implications of market integration. This phenomenon has a specific dimension in Borneo, where local systems regulating access to land and natural resources are complex and well-established, though now under attack from both without and within (Geddes 1954, Cleary 1997, Doolittle 2003, Ngidang 2005).

The interplay between the state, the corporate sector and local communities is a worldwide and intensifying process. Regional studies and case studies gathered here are therefore concerned with contestation and negotiation over natural resources in acknowledging underlying forces, whether state initiatives, cultural practices, or local and global knowledge are at play (Watts 2000: 259; Edelman and Haugerud 2004: 30). In particular, the dual process of the gradual withdrawal of the state as the main driver of agricultural expansion and its replacement by large corporations appears exceptionally dynamic in Borneo. Investigating its logic and its likely implications, including those well beyond the shores of Borneo, are essential goals of this book.

The concepts running through the four chapters that follow thus provide, first, windows to better identify the prominent features and implications of agricultural expansion and agrarian transformation on Borneo Island, and, second, lessons and questions about the future. Constituting the core of the book, these chapters deal in turn with Sarawak, Sabah and Kalimantan. The first two (Chapters 3 and 4) concern Sarawak, currently the object of exceptionally rapid agricultural expansion. In Chapter 3, Rob Cramb reviews the historical interlinking of the processes of agricultural intensification and expansion in Sarawak's agrarian transition, contributing new insights on current theoretical and policy debates, including through an elaborate discussion of Boserup's (1965) emphasis on the link between land-use systems and land tenure. In particular, he demonstrates how Sarawak has in fact gone through three major and partially overlapping transition stages, the current dominant one being the transition from smallholdings to estates. Cramb emphasizes how much this is transforming both the agricultural landscape and the labour scene, with legal and illegal Indonesian workers providing most of the labour on the rapidly expanding private oil palm estates. In Chapter 4, through two case studies, Jean-François Bissonnette provides an analysis of the current dynamics of land struggles in Sarawak. These include the local populations' capacity to resist—sometimes with the help of NGOs—encroachment by large corporations, as well as the support that the latter obtain from the state. However, resisting encroachment does not prevent a substantial number of households from taking up oil palm cultivation on their own. In Chapter 5, Stéphane Bernard and Jean-François Bissonnette emphasize the longer tradition of oil palm plantation expansion in Sabah and its historical significance as well as contemporary implications for the economy and the environment. This includes addressing the issue of foreign workers in the plantation sector as well as the apparently systematic commoditization of forestland, largely through the intervention of the local state. In the substantive sixth chapter, Lesley Potter addresses complex social and environmental problems deriving from agricultural change in Kalimantan. She shows how often massive wastage of forestland has occurred, particularly under the Suharto regime, and how, since the regime's downfall, the local provincial authorities have continued to try and apply grandiose plans of exploiting or overexploiting the resource frontier, here again mostly through oil palm expansion. Reviewing recent changes in livelihood strategies and state development programmes, she stresses the dynamic links between the related but at the same time distinct transition processes occurring in the four provinces that make up Kalimantan.

Together, these chapters provide both an overview and an in-depth study of local, state or national and global socio-environmental processes at play in contemporary Borneo. They illustrate the intensity and complexity of changes associated with agricultural expansion and intensification as well as local responses. Whether this expansion is likely to continue in Borneo, spread or even be transferred beyond its shores is an issue to which we will return in the Conclusion.

Notes

1. There is a daily bus service between Kuching and Pontianak.
2. For Kalimantan ethnic groups, see Table 6.2 in Chapter 6.
3. Although dated 2000, this map was based on "an image mosaic … generated from a full set of S10-products for the following months: March to September 1998, March to November 1999 and—for a few selected areas—January to March 2000" (GVM-TREES Project 2002).
4. In its November 2008 edition, *National Geographic* published an article titled "Borneo's Moment of Truth" (White and Klum 2008). Besides spectacular photos of the impact of oil palm expansion, it contains a very detailed double-page map titled "Vanishing Forests" and cites a long list of sources, some very recent. Yet, its land-use components are almost totally identical to those represented in the GVM-TREES Project *2000 Forest Cover Map of Insular Southeast Asia*.
5. In Figure 2.11, for 1957, "forested area" includes the following land-use categories listed in the *World Atlas of Agriculture*: mangrove forest, swamp forest, inland swamp forest, lowland forest, dipterocarp forest, Kerang forest and mountain forest. For 2000, it includes the following land-use categories listed in the *2000 Forest Cover Map of Insular Southeast Asia*: mangrove forest, swamp forest, evergreen lowland forest and evergreen montane forest.
6. In meteorological terms, "eye of the storm" actually refers to the quiet zone at the centre of a revolving storm. Here, we use this expression to refer to what it means in popular terms, namely, the centre of the storm.

References

Achard, F., H.J. Stibig, H. Eva and P. Mayaux, "Tropical Forest Cover Monitoring in the Humid Tropics—TREES Project", *Tropical Ecology* 43, no. 1 (2002): 9–20.

Achwan, R., *Microfinance Institutions, Social Capital and Peace Building: Evidence from West Kalimantan, Indonesia*. Jakarta: University of Indonesia, Center for Research on Inter-group Relations and Conflict Resolution, 2007.

Achwan, R. *et al.*, *Overcoming Violent Conflict: Peace and Development Analysis in West Kalimantan, Central Kalimantan and Madura*. Jakarta: CPRU-UNDP, LabSosio and BAPPENAS, 2005.

Appell, G., "Community Resources in Borneo: Failure of the Concept of Common Property and Its Implications for the Conservation of Forest Resources and the Protection of Indigenous Land Rights", *Bulletin Series No. 98*, Yale School of Forestry and Environmental Studies, 1995.

Ardhana, K., J. Langub and D. Chew, "Borders of Kinship and Ethnicity: Cross-Border Relations between the Kelalan Valley, Sarawak, and the Bawan Valley, East Kalimantan", *Borneo Research Bulletin* 35, no. 1 (2004): 144–79.

Bernard, S., "Palm Oil Expansion, Bio-fuel Production and Biodiversity Protection in Malaysia: Local Impacts of a Global 'Green Energy' Production Strategy". Paper presented at the IGU Brisbane Conference, Queensland University of Technology, July 2006.

Brookfield, H. and Y. Byron, "Deforestation and Timber Extraction in Borneo and the Malay Peninsula: The Record since 1965", *Global Environmental Change: Human and Policy Dimensions* 1, no. 1 (1990): 42–56.

Brookfield, H.C., Y. Byron and L. Potter, *In Place of the Forest: Environmental and Socio-economic Transformation in Borneo and the Eastern Malay Peninsula*. Tokyo and New York: United Nations University Press, 1995.

Brosius, P., "Transcripts, Divergent Paths: Resistance and Acquiescence to Logging in Sarawak, East Malaysia", *Comparative Studies in Society and History* 39 (1997): 468–510.

Casson, A., "Decentralisation, Forests and Estate Crops in Kutai Barat District, East Kalimantan", in *State, Communities and Forests in Contemporary Borneo*, Asia-Pacific Environment Monograph 1, ed. F. Majid-Cooke. Canberra: ANU E Press, 2006, pp. 65–86.

Cleary, M., "The Regulation of Land in North Borneo under the Chartered Company", *Borneo Review* 8, no. 1 (1997): 45–61.

Cleary, M. and P. Eaton, *Borneo: Change and Development*. New York: Oxford University Press, 1992.

Cleary, M. and F.J. Lian, "On the Geography of Borneo", *Progress in Human Geography* 15, no. 2 (1991): 163–77.

Colchester, M., "Pirates, Squatters and Poachers: The Political Ecology of Dispossession of the Native Peoples of Sarawak", *Global Ecology and Biogeography Letters* 3 (1993): 158–79.

Colchester, M., N. Jiwan, Andiko, M. Sirait, A.Y. Firdaus, A. Surambo and H. Pane, *Promised Land. Palm Oil and Land Acquisition in Indonesia: Implications for Local Communities and Indigenous Peoples*. Moreton-in-Marsh: Forest Peoples Programme, 2006.

Cramb, R.A., "The Importance of Secondary Crops in Iban Hill Rice Farming", *Sarawak Museum Journal* 34, no. 55 (1985): 37–45.

———, "Shifting Cultivation and Sustainable Agriculture in East Malaysia: A Longitudinal Case Study", *Agricultural Systems* 42 (1993): 209–26.

———, *Land and Longhouse, Agrarian Transformation in the Uplands of Sarawak*. Copenhagen: NIAS Press, 2007.

Dambul, R. and P. Jones, "Regional and Temporal Climatic Classification for Borneo", *Geografia* 5, no. 1 (2008): 1–25.

Dandot, W.B., "Land Development Programme as a Development Strategy in Bidayuh Areas", *Jurnal AZAM* 8, no. 1 (1992): 141–61.

Dauvergne, P., *Shadows in the Forest: Japan and the Politics of Timber in Southeast Asia*. Boston: Massachusetts Institute of Technology, 1997.

De Koninck, R., "Enjeux et stratégies spatiales de l'État en Malaysia", Hérodote 22 (1981a): 84–115.

_____, "Travail, espace, pouvoir dans les rizières du Kedah: Réflexions sur la dépossession d'un territoire", Cahiers de géographie du Québec 66 (1981b): 441–51.

_____, "Work, Space and Power in the Rice Fields of Kedah: Reflections on the Dispossession of a Territory", in The Southeast Asian Environment, ed. D.R. Webster. Ottawa: University of Ottawa Press, 1983, pp. 83–97.

_____, Malaysia: La dualité territoriale. Paris: Belin, 2007.

Doolittle, A., "From Village Land to 'Native Reserve': Changes in Property Rights in Sabah, Malaysia, 1950–1996", Human Ecology 29, no. 1 (2001): 69–98.

_____, "Colliding Discourses: Western Land Laws and Native Customary Rights in North Borneo, 1881–1918", Journal of Southeast Asian Studies 34, no. 1 (2003): 97–126.

Dove, M.R., "Peasant versus Government Perception and Use of the Environment: A Case-Study of Banjarese Ecology and River Basin Development in South Kalimantan", Journal of Southeast Asian Studies 17, no. 1 (1986): 113–36.

_____, "Smallholder Rubber and Swidden Agriculture in Borneo: A Sustainable Adaptation to the Ecology and Economy of the Tropical Forest", Economic Botany 47, no. 2 (1993): 136–47.

Drummond, I. and D. Taylor, "Forest Utilisation in Sarawak, Malaysia: A Case of Sustaining the Unsustainable", Singapore Journal of Tropical Geography 18, no. 2 (1997): 141–62.

Durand, F., "La question foncière aux Indes Néerlandaises, enjeux économiques et luttes politiques (1619–1942)", Archipel 58 (1999): 73–88.

Edelman, M. and A. Haugerud, The Anthropology of Development and Globalization: From Classical Political Economy to Contemporary Neoliberalism. Blackwell Anthologies in Social and Cultural Anthropology, 2004.

Eghenter, C., "Histories of Conservation or Exploitation? Case Studies from the Interior of Indonesian Borneo", in Histories of the Borneo Environment: Economic, Political and Social Dimensions of Change and Continuity, ed. R.L. Wadley. Leiden: KITLV Press, 2005, pp. 87–107.

Fisher, C.A., South-east Asia: A Social, Economic and Political Geography, 2nd edition. London: Methuen, 1966.

Freeman, D., Report on the Iban. London: The Athlone Press, 1970.

Geddes, W., "Land Tenure of the Land Dayaks", Sarawak Museum Journal 6 (1954): 42–51.

Gupta, A., Ecology and Development in the Third World, 2nd edition. London: Routledge, 1998.

_____, ed., The Physical Geography of Southeast Asia. New York: Oxford University Press, 2005.

Hong, E., Natives of Sarawak: Survival in Borneo's Vanishing Forests. Pulau Pinang: Institut Masyarakat, 1989.

Ishikawa N., A. Tanabe and P. Abinales, eds., Dislocating Nation-States: Globalization in Asia and Africa. Kyoto: Kyoto University Press; Melbourne: Trans Pacific Press, 2005.

Jackson, J.C., *Sarawak: A Geographical Survey of a Developing State*. London: University of London Press, 1968.

Jomo, K.S., Y.T. Chang and K.J. Khoo, *Deforesting Malaysia: The Political Economy and Social Ecology of Agricultural Expansion and Commercial Logging*. London: Zed Books, 2004.

Jones, L.W., "La population de trois États de Bornéo: Bornéo du Nord, Sarawak, Brunei", *Population* 19, no. 2 (1964): 325–34.

Kalimantan Review, Edisi Khusus Tahun XII. Sarawak, 2003.

King, V.T., "*Politik pembangunan*: The Political Economy of Rainforest Exploitation and Development in Sarawak, East Malaysia", *Global Ecology and Biogeography Letters* 3 (1993): 235–44.

Lee Yong Leng, *North Borneo: A Study in Settlement Geography*. Singapore: Eastern Universities Press, 1965.

Levang, P., N. Buyse, S. Sitorus and E. Dounias, "Impact de la décentralisation sur la gestion des ressources forestières en Indonésie: Études de cas à Kalimantan-Est", *Anthropologie et Sociétés* 29, no. 1 (2005): 81–102.

MacKinnon, K. and E. Sumardja, "Forests for the Future: Conservation in Kalimantan", in *Borneo in Transition: People, Forests, Conservation, and Development*, 2nd edition, ed. C. Padoch and N.L. Peluso. Kuala Lumpur: Oxford University Press, 2003, pp. 71–92.

Majid-Cooke, F., "The Politics of Sustained Yield Forest Management in Malaysia: Constructing the Boundaries of Time, Control and Consent", *Geoforum* 26, no. 4 (1996): 445–58.

————, "Vulnerability, Control and Oil Palm in Sarawak: Globalization and a New Era?" *Development and Change* 33, no. 2 (2002): 189–211.

————, "Maps and Counter Maps: Globalised Imaginings and Local Realities of Sarawak's Plantation Agriculture", *Journal of Southeast Asian Studies* 34, no. 2 (2003): 265–84.

————, ed., *State, Communities and Forests in Contemporary Borneo*. Canberra: ANU E Press, 2006.

McCarthy, J., "The Changing Regime: Forest Property and Reformasi in Indonesia", *Development and Change* 31, no. 1 (2000): 91–129.

————, "Changing to Gray: Decentralization and the Emergence of Volatile Socio-Legal Configurations in Central Kalimantan, Indonesia", *World Development* 32, no. 7 (2004): 1199–223.

McMorrow, J. and A.M. Talip, "Decline of Forest Area in Sabah, Malaysia: Relationship to State Policies, Land Code and Land Capability", *Global Environmental Change* 11 (2001): 217–30.

Morrison, P.S., "Transitions in Rural Sarawak: Off-Farm Employment in the Kemena Basin", *Pacific Viewpoint* 34, no. 1 (1993): 45–68.

Ngidang, D., "Contradictions in Land Development Schemes: The Case of Joint Ventures in Sarawak, Malaysia", *Asia Pacific Viewpoint* 43, no. 2 (2002): 157–80.

————, "Deconstruction and Reconstruction of Native Customary Land Tenure in Sarawak", *Southeast Asian Studies* 43, no. 1 (2005): 47–75.

Padoch, C., "Labor Efficiency and Intensity of Land Use in Rice Production: An Example from Kalimantan", *Human Ecology* 13, no. 3 (1985): 271–89.

Padoch, C. and N.L. Peluso, eds., *Borneo in Transition: People, Forests, Conservation, and Development*, 2nd edition. Kuala Lumpur: Oxford University Press, 2003.

Potter, L., "Forest Degradation, Deforestation, and Reforestation in Kalimantan: Towards a Sustainable Land Use?" in *Borneo in Transition: People, Forests, Conservation and Development*, 2nd edition, ed. C. Padoch and N.L. Peluso. Kuala Lumpur: Oxford University Press, 2003.

_____, "Commodifying, Consuming and Converting Kalimantan's Forests, 1950–2002", in *Muddied Waters: Historical and Contemporary Perspectives on Management of Forests and Fisheries in Island Southeast Asia*, ed. P. Boomgaard, D. Henley and M. Osseweijer. Leiden: KITLV Press, 2005, pp. 373–99.

Potter, L., H. Brookfield and Y. Byron, "The Eastern Sundaland Region of South-east Asia", in *Regions at Risk: Comparisons of Threatened Environments*, ed. J.X. Kasperson, R.E. Kasperson and B.L. Turner II. Tokyo, New York and Paris: United Nations University Press, 1995.

Raffestin, C. and M. Bresso, *Travail, espace, pouvoir*. Lausanne: l'Âge d'Homme, 1976.

Raharto, A., "Indonesian Labour Migration", *International Journal on Multicultural Societies* 9, no. 2 (2007): 219–35.

Rhee, S., "De Facto Decentralization and Community Conflicts in East Kalimantan, Indonesia: Explanations from Local History and Implications for Community Forestry", in *The Political Ecology of Tropical Forest in Southeast Asia: Historical Perspectives*, ed. Lye Tuck-Po, W. de Jong and A. Ken-ichi. Kyoto: Kyoto University Press, 2003, pp. 152–76.

Rimmer, P.J., "The Spatial Impact of Innovation in International Sea and Air Transport since 1960", in *Southeast Asia Transformed: A Geography of Change*, ed. Chia Lin Sien. Singapore: Institute of Southeast Asian Studies, 2003, pp. 296–316.

Rosser, A., K. Roesad and D. Edwin, "Indonesia: The Politics of Inclusion", *Journal of Contemporary Asia* 35, no. 1 (2005): 53–78.

Rousseau, J., *Central Borneo: Ethnic Identity and Social Life in a Stratified Society*. Oxford: Clarendon Press, 1990.

Sellato, B., *Hornbill and Dragon: Arts and Culture of Borneo*, 2nd edition. Singapore: Sun Tree Publishing, 1992.

Sercombe, P.G. and B. Sellato, eds., *Beyond the Green Myth: Borneo's Hunter-Gatherers in the 21st Century*. Copenhagen: NIAS Press, 2007.

Siegert, F. and A.A. Hoffmann, "The 1998 Forest Fires in East Kalimantan (Indonesia): A Quantitative Evaluation Using High Resolution, Multitemporal ERS-2 SAR Images and NOAA-AVHHR Hotspot Data", *Remote Sensing of Environment* 72, no. 1 (2000): 64–77.

Sunderlin, W., "Between Danger and Opportunity: Indonesia and Forests in an Era of Economic Crisis and Political Change", *Society & Natural Resources* 12, no. 6 (1999): 559–70.

Suryadinata, L., E.N. Arifin and A. Ananta, *Indonesia's Population: Ethnicity and Religion in a Changing Political Landscape*. Singapore: Institute of Southeast Asian Studies, 2003.

Vandergeest, P. and N.L. Peluso, "Territorialization and State Power in Thailand", *Theory and Society* 24, no. 3 (1995): 385–426.

Wadley, R. and O. Mertz, "Pepper in a Time of Crisis: Smallholder Buffering Strategies in Sarawak, Malaysia and West Kalimantan, Indonesia", *Agricultural Systems* 85, no. 3 (2005): 289–305.

Warren, C., "Mapping Common Futures: Customary Communities, NGOs and the State in Indonesia's Reform Era", *Development and Change* 36, no. 1 (2005): 47–73.

Watts, M., *People, Plants, and Justice: The Politics of Nature Conservation*. New York: Columbia University Press, 2000.

White, M. and M. Klum, "Borneo's Moment of Truth", *National Geographic* (Nov. 2008): 34–63.

3

Agrarian Transitions in Sarawak: Intensification and Expansion Reconsidered

R.A. Cramb[1]
The University of Queensland

"We want population to turn our wasteland into shape and create bustle and industry ... [We want] to see the jungle falling left and right and people settled over what are now lonely wastes and turning them into cultivated lands." (Charles Brooke, Second Rajah of Sarawak, 1867)[2]

"My vision for the next twenty years is to see modern agricultural development along the major trunk road with rows of plantations and villages well organised in centrally managed estates with a stake of their own in them." (Abdul Taib Mahmud, Chief Minister of Sarawak, 1984)[3]

Introduction

Among the central processes said to be involved in the agrarian transition in Southeast Asia are the apparently contrasting phenomena of agricultural intensification and territorial expansion (De Koninck 2004). Agricultural intensification is usually seen as increased productivity of existing croplands through use of high-yielding varieties of foodgrains, greater application of inputs (fertilizers, pesticides and irrigation), and the employment of more labour (particularly through double cropping). This is typically associated with increased links with markets and growing inequality in access to land, employment and income (Gibbons *et al.* 1980, Hart *et al.* 1989). The so-called green revolution in lowland rice production in Southeast Asia, typified by the massive Muda Irrigation Scheme in Peninsular Malaysia, is the paradigmatic case of such agricultural intensification (Scott 1985, Johnson

44

2000). Agricultural expansion, by contrast, is seen as involving the movement of populations into new territory, that is, land pioneering on the forest margin, pushing back the forests and converting them to permanent agricultural land (Pelzer 1948, De Koninck and McTaggart 1987, Angelsen 2007). Such land settlement is typically (though not necessarily) state-sponsored and centrally managed, and has often involved extensive planting of commercial tree crops such as rubber and oil palm. The Federal Land Development Authority (FELDA) schemes in Peninsular Malaysia are often cited as the prime example of this form of agricultural development (Sutton 1989).

The apparent absence of any green revolution in rice cultivation in Sarawak and the speed and extent of the oil palm boom that has swept across the rural landscape lend credence to the notion that expansion of the agricultural frontier is the key process in Sarawak's agrarian transition, converting forestland to modern plantation agriculture on an unprecedented scale. The area under oil palm has increased 25-fold over the past two and a half decades, from 23,000 hectares in 1980, mostly in government schemes, to more than 590,000 hectares in 2006, mostly in private estates, closing in on an official target of 1 million hectares by 2010 (Figure 3.1). By 2005, oil palm accounted for about 57 per cent of the area under agricultural crops (Table 3.1). A concomitant of this perception that large-scale agricultural expansion is the driving force in Sarawak is the belief that "idle" or "waste" lands are being drawn into the development process, thus creating employment and income, reducing rural poverty, and bringing economically backward people into the "mainstream of modernization".

In this chapter, I draw on the seminal work by Boserup (1965) to argue that the processes of agricultural intensification and expansion are, at least in the Sarawak context, much less distinct than might at first appear, with important implications for both an understanding of the agrarian transition and the formulation of agrarian policy. Rather than viewing agrarian change in Sarawak as a monotonically expanding (capitalist) plantation frontier, I argue for a closer, more fine-grained historical and geographical perspective that reveals successive waves of expansion and contraction, within which are found pockets of intensification and "disintensification". The latter involve a diverse array of actors, including independent, village-based and "managed" smallholders; various government departments and agencies; donor organizations; and private agribusiness and timber firms. Moreover, this complex process of land-use change has been interwoven with successive and often competing claims to land and forest resources, resulting in an "institutional layering" of customary, statutory and de facto property rights, ensuring that the agrarian transition in Sarawak is highly problematic and contested (Cramb and Wills 1990, Ngidang 1997, Majid-Cooke 2006, Cramb 2007).[4]

Figure 3.1 Sarawak. Trends in planted area of major crops, 1985-2005

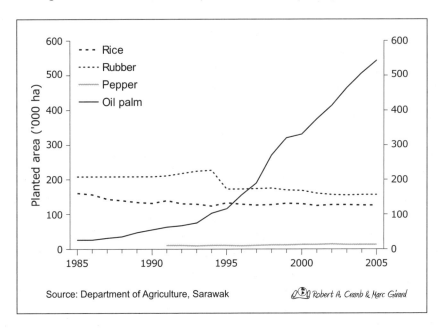

Source: Department of Agriculture, Sarawak   Robert A. Cramb & Marc Girard

Table 3.1 Sarawak. Agricultural land use, 2005

Crop	Area (ha)	% of total area
Rice	127,220	13.2
Rubber	156,839	16.3
Pepper	12,673	1.3
Sago	51,763	5.4
Coconut	24,593	2.6
Oil palm	543,399	56.6
Cocoa	4,898	0.5
Fruit trees	34,210	3.6
Vegetables	3,112	0.3
Other	2,193	0.2
Total	960,900	100.0

Source: Department of Agriculture, Sarawak.

From this longer-term, spatial perspective it is possible, in fact, to discern three historic agrarian transitions in Sarawak—the transition to shifting cultivation, the transition to smallholder cash crops, and the transition to large-scale plantation agriculture—resulting in a series of partially overlapping and mutually determining (or "imbricated") socio-ecological landscapes.[5] That is, not only has each agrarian transition led to the transformation of the preceding landscape (including the substratum of initial ecological conditions in Sarawak), but the characteristics of that landscape have strongly influenced the contours of succeeding landscapes.[6] These mutual influences are repeatedly illustrated in the three main sections of the chapter, following an elaboration and extension of Boserup's contribution to understanding these processes.

Boserup in Borneo

Classical economists introduced the theoretical distinction between the intensive and extensive margins of cultivation that underlies the concepts of agricultural intensification and expansion (Barlowe 1986, Van Kooten 1993, Blaug 1997). They argued, first, that "as the price of agricultural output rises, production will expand onto marginal land", resulting in "rent accruing at the extensive margin" (Van Kooten 1993: 22). By "marginal land" Ricardo meant land of lower inherent quality or fertility, while Von Thunen emphasized increasing distance from a central market as the primary determinant of the land's marginality (Barlowe 1986, Angelsen 2007). For both writers, land beyond the extensive margin was "waste" or "wilderness". At the same time, Ricardo (and also Marx) argued that rising prices induced the use of more labour and material inputs on *existing* croplands, thus increasing production and rent at the *intensive* margin (Barlowe 1986, Van Kooten 1993). As Black (1929) long ago pointed out, the two responses, though distinct, were seen as part and parcel of the same economic process of adjusting land use to market conditions. That is, a rise in the price of agricultural output due to increasing demand relative to supply would be expected to result simultaneously in both intensification and expansion of agricultural land use, whether at the farm or regional scale.

This neat distinction between the intensive and extensive margins was challenged by Ester Boserup (1965) in her influential essay on the economics of agrarian change under population pressure:

> Why this approach is unsuitable for a general theory of agricultural development is most easily understood if it is remembered that many types of primitive agriculture make no use of permanent fields, but shift cultivation from plot to plot ... [It follows] ... that in primitive types of agriculture there

is no sharp distinction between cultivated and uncultivated land, and that it is impossible, likewise, to distinguish clearly between the creation of new fields and the change of methods in existing fields. (Boserup 1965: 12–3)

Consequently, in Boserup's analysis of agrarian change, "the very distinction between fields and uncultivated land is discarded and instead emphasis is placed on the frequency with which the land is cropped" (Boserup 1965: 13). Cropping frequency thus becomes the principal indicator of land-use intensity. This gives rise to her well-known sequence of land-use intensification in response to increasing population pressure: from forest-fallow to bush-fallow to short-fallow to annual cropping to multi-cropping, with changes in cultivation technology along the way. Of course, regarding all these changes as "intensification" assumes the entire landscape has already been brought into cultivation (however extensive) at some point. In later contributions, Boserup (1976, 1981) includes "gathering" in the sequence of food production systems, involving no period of cultivation at all. Hence, so long as a forest landscape is occupied by bands of hunter-gatherers (as Borneo has been for millennia), the process of land-use change with population growth can be viewed as one of waves of intensification rather than expansion.[7]

Boserup is often criticized for positing an oversimplified, unilinear sequence of land-use change. However, she emphasizes that cultivation systems of differing intensity may coexist within a given territory, perhaps for centuries (Boserup 1965: 62–3). Other writers have invoked Von Thunen to explain the "rings" of land use around a village centre in a developing economy where transportation is largely dependent on human effort (Chisholm 1979; Ruthenberg 1980: 75–9). Thus, permanent agricultural plots may be found close to the centre of settlement while plots on the outer margins are still within a long-fallow system and/or used for foraging. Boserup (1965: 62–3) also explicitly allows for the "regression in agricultural techniques" that occurs during periods of declining population pressure. Thus, intensive farmers who are relocated to regions of lower population density typically adopt more extensive food production systems (e.g., Javanese transmigrants in upland areas of South Kalimantan [Masyhuri and Cramb 1995]). Despite Boserup's use of pejorative terms such as "primitive" and "regression", the thrust of her argument is that extensive systems of land use such as forest-fallow farming have an economic and ecological rationale and are not an indication of cultural backwardness or irrational resistance to change.

Boserup's (1965, 1976) theory explicitly minimizes the significance of environmental constraints to intensification, a point debated at length by later writers (Brookfield 1972, 1984; Chin 1977; Tiffen and Mortimore 1994). Pingali and Binswanger (1987) have extended her analysis by taking into account the characteristics of different environments along a toposequence from uplands to lowlands. In addition, they allow for not only the growth of population but also migration between zones. Recognizing the variability in what Brookfield (1972) terms environmental "elasticity" means that the full intensification sequence proposed by Boserup, with progressively longer cropping periods and shorter fallow periods, is not feasible in much of the uplands of Borneo without causing serious land degradation, increasing poverty, and outmigration (Chin 1977, Cramb 2005).

Raintree and Warner (1986) have also elaborated on Boserup's theory of intensification, taking account of such environmental constraints. They outline a variety of "agroforestry" pathways that open up at different stages, such as enriched fallows in the forest and bush-fallow stages and alley cropping in the short-fallow and annual cropping stages. In particular, they highlight that tree crops provide an alternative intensification pathway, even at relatively low population densities. Thus, with population growth and the improvement of rural infrastructure, shifting cultivators in Borneo have frequently been motivated to incorporate tree crops such as rubber, coffee and cocoa in their farming systems rather than push shifting cultivation beyond what Blaikie and Brookfield (1987) term the "ecological margin". This necessarily means moving beyond largely subsistence production to at least partial engagement with local and global markets (Cramb 1988a, Dove 1993, Tiffen and Mortimore 1994).[8]

According to Barlow and Jayasuriya (1986), the development of small-holder tree crop cultivation has proceeded through three historical phases. The first is "emergence from subsistence", when subsistence production is supplemented by a plantation crop. Simple, labour-intensive, tree-crop technologies are rapidly adopted by smallholders, typically through diffusion from estates. This is followed by the "agricultural transformation" stage, when smallholder farming becomes largely commercialized and new high-yielding tree-crop technologies are progressively adopted. Finally, the phase of "extended structural change" is characterized by the increasing significance of the industry and service sectors in the economy, rendering smallholder tree crops less profitable due to the rising cost of land and labour. Tomich *et al.* (1995) identify a significant demographic turning point in this period of structural change, namely, when the absolute size of the agricultural workforce peaks and begins to fall. Barlow (1997) shows that the development of smallholder rubber in Malaysia has experienced all three phases, whereas in countries such as Indonesia rubber is in the (late) agricultural transformation stage.

Thus, the incorporation of tree crops in shifting cultivation systems in Borneo, while seen as an "expansion" of the area under the crop in question, is in Boserup's terms an "intensification" of the land-use system within a given territory. The subsequent adoption of higher-yielding tree-crop technologies constitutes a further round of intensification. However, the phase of "extended structural change", involving the loss of labour from rural pursuits, if not from rural areas altogether, creates pressures for the "disintensification" of the entire land-use system, with both food and tree-crop production scaled back. This relates to the problem of "idle land", which emerged as a policy issue in Peninsular Malaysia in the 1980s and 1990s (Vincent and Mohamed Ali 1997, Angelsen 2007), until partly offset by the influx of impoverished Indonesian plantation labour.

A major strand in Boserup's (1965) theory of agrarian change that has been less discussed is her emphasis on the link between land-use systems and land tenure. She argues that in forest-fallow systems households possess a general cultivation right to village lands, which translates into a temporary use right to any particular field the household clears. This use right reverts to the village or lineage when the field is fallowed. As land-use intensity increases and the fallow period is shortened, the specific use right becomes more valuable and permanent, while the general access right diminishes in importance. Binswanger and McIntire (1987) further theorize that the introduction of commercial crops, especially tree crops, accelerates the emergence of specific rights. As these become transferable within and beyond the village, land acquires a market value, and also a collateral value, inducing expansion in the supply of credit. These processes lead to inequality in landholdings and the emergence of landlessness and tenancy. Such outcomes are widely observable in the more open, economically differentiated villages of lowland Southeast Asia but can also occur in upland regions, particularly where migration and resettlement have undermined customary institutions of community control (Li 2002, Cramb and Culasero 2003).

Where community governance has been overlain or superseded by state control of land tenure, outcomes frequently differ from Boserup's evolutionary model. Elson (1997) shows how indigenous and colonial states often intervened to maintain communal lands in accessible, densely populated villages in order to control the allocation of valuable cultivation rights, while in less densely populated frontier settlements, removed from state control, individual household rights generally prevailed. In the postcolonial era, well-developed individual rights of upland farmers have been overridden in communist states such as Vietnam and Laos, while common property rights to village forest reserves have been undermined by land laws and policies in parts of Indonesia and Malaysia. Such interventions in customary tenure systems illustrate the use by modern states of territorial strategies to establish control over natural resources and the people who use them (Vandergeest and Peluso 1995, Peluso and Vandergeest 2001). This process of internal territorialization is about "excluding or including people within particular geographic boundaries, and about controlling what people do and their access to natural resources within those boundaries" (Vandergeest and Peluso 1995: 388).

Thus, the concept of agricultural expansion at the extensive margin is not merely an economic concept but fundamentally a political one (Blaikie and Brookfield 1987, De Koninck 2000, Angelsen 2007).[9] If lands beyond the margin of permanent, commercial agriculture are viewed as "idle lands" or (to use Rajah Charles Brooke's phrase) "lonely wastes", they are simultaneously rendered "state spaces", available to be appropriated by state-sanctioned actors for purportedly more productive or beneficial uses, regardless of prior occupation and usage (Peluso 1995, De Koninck 1996, Scott 1998, Majid-Cooke 2006). Questions of who occupies the "wastelands" and what is the legitimacy of their claims to tenure are thus obscured. Even Boserup, in setting up her critique of the extensive margin, blurs the underlying politics: "The classical economists were writing at a time when the *almost empty lands* of the Western Hemisphere were gradually taken under cultivation by European settlers, and it was therefore natural that they should stress the importance of the *reserves of virgin land* ..." (1965: 12; emphasis added). A similar "Ricardian slip" is evident in a recent news article justifying the controversial Murum Dam project in the interior of Sarawak, where the lands to be flooded are described as "virtually empty", thus rendering insignificant the presence of Penan settlements in that watershed. This obscuring of the crucial links between expansion, occupation, intensification and exclusion is evident in the process of agrarian transition under way in Sarawak, as elsewhere in Borneo (Cleary and Eaton 1992, Padoch and Peluso 2003).

The Transition to Shifting Cultivation

Sago, Rice and People

The first agrarian transition in Sarawak was from a foraging-cum-horticultural economy centred on the sago palm to an economy based on the shifting cultivation of rice. The earliest Austronesian occupants of Sarawak's rainforests were most likely small bands of hunter-gatherers who moved through extensive but defined territories utilizing and managing a wide variety of naturally occurring plant and animal resources (King 1993).[10] Hill sago (*Eugeissona utilis* and other species) was harvested on a cyclical basis as a major source of subsistence, necessitating frequent movement between groves of sago to allow regeneration (Brosius 1986, Sellato 1994). Consequently, though food supplies were abundant, population densities were necessarily low. At some point, swamp sago (*Metroxylon sagu*), originating in the Moluccas or New Guinea, was introduced to the lowlands of Borneo, permitting more intensive cultivation (Morris 1991, Ellen 2004). Barton and Denham (in press) argue that pre-rice agricultural systems in Borneo were largely based on "vegeculture"—the cultivation or husbanding of long-lived plants by vegetative propagation—including swamp and hill sago and also root crops such as taro (*Colocasia esculenta*). This is consistent with King's view that in the first millennium AD, "horticulture was still the dominant form of agriculture in Borneo" and that "the widespread adoption of shifting cultivation of rice ... had to await the more general use and manufacture of iron tools in Borneo ... [which] made rainforest clearance less arduous" (1993: 99, 102). The development of ironworking and the widespread use of iron tools seem to have begun later in the millennium, first in coastal sites such as Santubong at the mouth of the Sarawak River, and spreading slowly into the interior (Cleary and Eaton 1992: 28–9; King 1993: 99–102).

Certainly by the middle of the second millennium AD, longhouse-dwelling Dayak tribes practising shifting cultivation of hill rice were on the move in Borneo. As populations took up shifting cultivation, they found a need to expand into surrounding forestlands to support a long forest-fallow cycle and maintain access to abundant forest resources.[11] Though this necessarily involved "pioneering", in the sense of felling old-growth forest, in general it did not involve abandoning lands already farmed (Padoch 1982a, 1982b).[12] Rather, subgroups would hive off and move away into new territory, leaving the original homelands still occupied.[13] The Bidayuh and Selakau of western Borneo expanded gradually northeastwards across the mountains into southwest Sarawak, and now make up 5 per cent of Sarawak's population (Geddes 1954a, Schneider 1978). The more numerous and aggressive Iban (now 33 per cent of the population) spread from the Kapuas basin in a predominantly northeasterly direction into the hilly midlands of Sarawak, settling first in the Lupar and Saribas basins from about the 16th century (Sandin 1956, 1967; Pringle 1970). As Heppell *et al.* write:

> On these migrations, the Iban displaced earlier settlers who were usually hunter/gatherers. The Iban occupied the forest through force of arms. Any groups occupying territory in their way were either annihilated or absorbed into Iban society as slaves (2005: 18).

Some hunter-gatherer groups did resist the Iban advance for an extended period, particularly in the Krian (Sandin 1967, Pringle 1970). Nevertheless, by the early 19th century (before the advent of the Brookes) the northeasterly expansion of the Iban had taken them into the Rejang watershed and beyond.

This brought them into conflict not only with foraging groups, but also with other groups of shifting cultivators, notably the Kayan and Kenyah, who had been expanding in a generally southwesterly direction from their homeland in the Apo Kayan into the upper reaches of the Baram and Rejang (Baluy and Baleh) Rivers, in the case of the Kayan, since the late 18th century (Pringle 1970; Rousseau 1974, 1978; Whittier 1978; Chin 1985; Ngo 2003).[14] The intervention of Charles Brooke in leading a punitive expedition of 15,000 Iban irregulars to the upper Rejang in the Great Kayan Expedition of 1863 helped decide this territorial contest in favour of the Iban (Pringle 1970: 130–4; Rousseau 1977).

The Iban and Kayan-Kenyah expansions into central and northern Sarawak cut a swathe through the territories of a pre-existing stratum of linguistically and culturally related groups who were still predominantly sago growers, from the Melanau along the central coast to other groups in the interior loosely classified as Kajang.[15] The Melanau-Kajang groups became separated from each other in what has been termed a "submerged complex" (King 1993).[16] The coastal Melanau (now 7 per cent of the population), though formerly longhouse dwellers, mostly adopted a Malay-Muslim lifestyle while retaining a distinctive emphasis on sago cultivation for both subsistence and export (Morris 1978, 1991), while the upriver Kajang aligned themselves with the dominant Kayan and Kenyah and increasingly adopted hill rice cultivation, though sago remained an important source of subsistence (Rousseau 1977, Nicolaisen 1986, Strickland 1986). The Kayan, Kenyah, Kajang and other interior groups such as the Kelabit and Lun Bawang of the northern highlands, collectively labelled Orang Ulu, make up 8 per cent of the population.[17]

Notwithstanding the spread of hill rice cultivation, foraging in both primary and secondary forest continued to be important. This was true both for the shifting cultivators themselves and for smaller, specialist groups such as the nomadic Penan who were often linked economically and politically to the nearby longhouse societies. The Penan are linguistically and culturally related to the Kenyah. Rather than constituting a remnant group left behind by more advanced agriculturalists, they continue to occupy a productive niche in the forest interior. Like the Kenyah before them, most Penan have taken up shifting cultivation in the past 50–100 years, though hunting and gathering (especially of wild sago) remain important to their subsistence (Brosius 1986, 1988; Langub 1988; King 1993: 167–70).

It is important to note that as well as practising shifting hill rice cultivation, the Iban, Bidayuh, Orang Ulu and other groups also cultivated swamp rice where the environment was suitable—in small inland valleys and along river margins. As Pringle points out, "Iban agriculture was not traditionally restricted to the cultivation of hill rice. There are downriver areas in the Second Division (Lupar and Saribas basins) where the Ibans have always cultivated what they call swamp rice (*padi paya*)" (1970: 26). Indeed, there is circumstantial evidence that their long-fallow system of hill rice (*padi bukit*) cultivation may have evolved from short-fallow swamp rice cultivation in the inland valleys of West Kalimantan (Seavoy 1973, Padoch *et al.* 1998, Cramb 2007). A survey of Iban farmers along the Oya River, spanning lowland, midland and upland environments, found a continuous progression from swamp rice (*padi paya*) to dry rice on flat land (*padi emperan*) to hill rice (*padi bukit*), with cropping intensity steadily falling along this toposequence (Cramb and Dian 1979a). Where pockets of bottomland are interspersed among low hills, the continuity between the two basic types of swamp rice and hill rice can be readily observed, often within the same farm. These pockets of swamp rice (some of them quite extensive, as in the lower Lupar) have provided a basis for significant intensification in recent decades, as discussed below.

The Sarawak Malays, descended from seagoing traders and Islamized Dayaks, and occupying settlements along the coastal and riparian fringe— predominantly in southwest Sarawak—also derived their livelihoods in part from swamp rice cultivation, but mainly from coastal fishing, foraging in the swamp forest, and small-scale riverine trade (Harrisson 1970, Pringle 1970, King 1993). Malay traders provided a historically important conduit for surplus rice and forest products from the shifting cultivators of the interior, until this role was progressively taken up by Chinese traders in the 19th century (Chew 1990). They also planted coconut palms, particularly between the Sarawak and Sadong River mouths, which became the basis of a minor local industry.[18] The Malays now make up 21 per cent of Sarawak's population but a much smaller proportion of the agricultural population.

Thus, the expansion of the shifting cultivation frontier across Sarawak over several centuries has involved major demographic and land-use changes, resulting in the transformation of the forest landscape and the intensification of food production systems within this landscape. Figure 3.2, compiled by Hatch (1982), shows the extent of shifting cultivation lands by the 1970s, including hill rice clearings, forest fallow, islands of communal forest reserve, and small plots of swamp rice, pepper, rubber and fruit trees. The area was estimated to be 31,780 square kilometres, or 26 per cent of Sarawak's total land surface (Hatch 1982). This shifting cultivation zone was still increasing marginally as some communities in remote districts continued to convert old-growth forest, but at an estimated annual rate of only 0.2 per cent (Cramb 1988b, 1990b; see also Hansen 2005). The long expansion phase of shifting cultivation in Sarawak had effectively ceased.

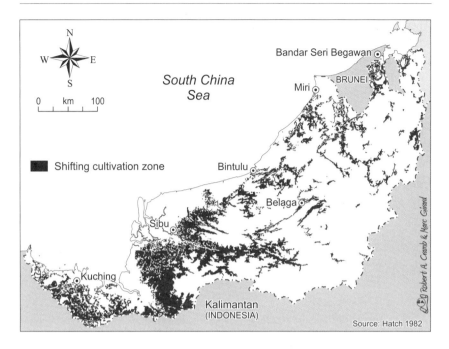

Figure 3.2 Sarawak. The shifting cultivation zone

Territory and Tenure

The expansion of shifting cultivation has also entailed significant changes in territorial control and land tenure. There is evidence that at least some bands of hunter-gatherers claimed extensive territories within which they moved and husbanded resources (termed *tana pengurip* in Penan), while in other cases bands moved freely into each other's range without concern for boundaries, though probably within the domain of the same tribe (Brosius 1986, 1988; King 1993; Sellato 1994; Lembat 1994; Langub 2007). However, at least among the Penan, individuals marked and preserved (*molong*) groves of sago, rattan and other forest resources for future use, and these claims were upheld by their own and other bands (Langub 1988, 2007). Certainly agriculturalists who cultivated sago, such as the Melanau and Kajang, claimed exclusive territorial rights on behalf of the longhouse community as well as household and even individual use rights to cultivated land and trees (Morris 1976, Nicolaisen 1986, Strickland 1986). The Iban, Bidayuh, Kayan and Kenyah shifting cultivators all established territorial control over the regions they moved into, displacing or absorbing hunter-gatherer groups in the process.[19]

They used armed force to establish and defend longhouse territorial boundaries, which included cultivated land and extensive forest reserves.

Within these territories the system of land tenure varied. While all recognized the general right of access of longhouse members, the mode of sharing specific use rights, first established through felling old-growth forest, took several forms. The Bidayuh system conformed to Boserup's depiction of a temporary use right reverting to a descent group (Geddes 1954b), while Iban, Kayan and Kenyah land tenure generally recognized permanent household cultivation rights, so long as the household remained in the longhouse community (Freeman 1970, Whittier 1978).[20] In some cases the community took over the annual allocation of cultivation rights, though this appeared to be a localized, historical adaptation to specific conditions (e.g., taking over the secondary forest of migrants) rather than reflecting a primordial communalism: see Rousseau (1974, 1987) on the Kayan and Cramb (1989) on the Iban. In all these cases, however, the customary law (*adat*) of the longhouse community provided effective governance of individual and group access to land and forest resources.

The territorialization strategies of the Sarawak state have been directed primarily at controlling the expansion of shifting cultivators and reserving forestland for logging and plantations. In the 19th century the Brookes were at first merely concerned to prevent the Iban in particular from getting too far away from government control. Ironwood markers (*pak*) were inserted to indicate the upriver boundaries of authorized settlement, but these boundaries proved difficult to enforce. By the 1920s and 1930s the Brooke government was endeavouring to establish permanent forest reserves, following the example of the Malayan Forest Service (Kaur 1995). A major internal report on Sarawak's administration condemned "our present policy of non-interference" and advocated that "legislation should be introduced to confine the operations of the shifting farmer to secondary forest". In addition, "having classified all forest areas, legislation should then be introduced to prohibit the felling of virgin forest except for permanent forms of agriculture" (Le Gros Clarke 1935: 31). Following the report's recommendations, the total area of forest reserves was increased from 1.2 per cent of Sarawak's land area in 1934 to 5.5 per cent in 1940.

The post-war British colonial government took sterner action to restrict shifting cultivation, increasing the area under forest reserves to 24 per cent by 1960 and introducing a total ban on the felling of primary forest for shifting cultivation, effectively from the introduction of the 1958 Land Code (though some local administrative approvals were given to clear primary forest in the 1960s). The Land Code rendered shifting cultivators mere "licensees of the state" in terms of land already cleared or otherwise occupied, and gave no recognition at all to territorial rights, undermining customary claims to the extensive forestlands reserved by shifting cultivation communities, as well as the hunting and gathering domains of the remaining nomadic groups (Brosius 1986, Cramb and Wills 1990). Amendments to the Land Code in the 1990s further undermined customary land rights (Ngidang 2005, Cramb 2007).

This legal defencelessness became apparent with the rapid expansion of mechanized, commercial logging in Sarawak's hill forests in the 1970s and 1980s, particularly in the upper Rejang and Baram watersheds. Figure 3.3 shows the area logged by 1992, though this has since expanded eastwards as far as the Indonesian border. The expansion of logging through and beyond the territories of upland communities (particularly Orang Ulu and Penan) sparked a prolonged and ongoing conflict between these communities, logging companies, and the forces of the state that has been well-documented elsewhere (State Planning Unit 1987; Hong 1987; WRM/SAM 1989; INSAN 1989; Colchester 1989; King 1993: 290–302; IDEAL 1999, 2000). The resistance of local people through blockades and legal avenues seemed only to strengthen the resolve of the Sarawak government to sweep them aside through draconian

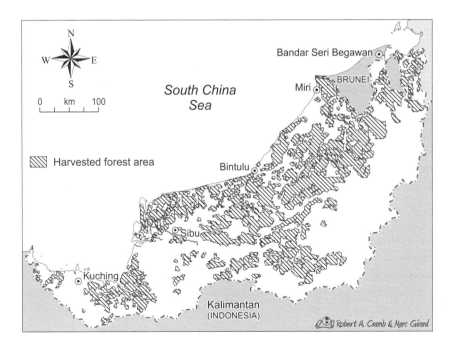

Figure 3.3 Sarawak. Logged area up to 1992

legislative changes and police action, as well as an ideological campaign that blamed shifting cultivators rather than logging for the forest destruction that was occurring (State Planning Unit 1987, Cramb 1988b, Chin 1989).[21]

There are two points to be made here in regard to agrarian transition. First, as a comparison of Figures 3.2 and 3.3 shows, the rapidly expanding logging frontier intensified the use of the forest well beyond the extensive margin of traditional long-fallow agriculture, both downriver into the swamp forests and upriver into previously inaccessible hill forests. Second, the expansion of logging prepared the way for and was intertwined with the transition from smallholder to plantation agriculture (discussed in a later section), both by providing access to forestland and by generating the surplus capital to invest in plantation development. Before considering this most recent and far-reaching transition, however, it is necessary to go back in time and review the emergence of commercial smallholder production, particularly among subsistence shifting cultivators.

The Transition from Shifting Cultivation to Smallholder Cash Crops

The second major agrarian transition in Sarawak, overlapping the first by a century or so, involved shifting cultivators inserting commercial tree and shrub crops into existing territories by converting a portion of their fallow lands under secondary forest into semi-permanent smallholdings or "gardens" (*kebun*), particularly of rubber (*Hevea brasiliensis*) and pepper (*Piper nigrum*). This occurred to varying degrees throughout the area shown in Figure 3.2. Although it involved an "expansion" of particular cash crops, it was essentially a (spatially concentrated) "intensification" of land use within the shifting cultivation zone. This development is consistent with the alternative intensification trajectories outlined by Raintree and Warner (1986), Barlow and Jayasuriya (1986), and Barlow (1997).

In particular, following Barlow's (1997) schema, there was a long phase of "emergence from subsistence" when swamp or hill rice production remained dominant and was merely supplemented by one or more cash crops. Since the 1970s this has been overtaken by an "agricultural transformation" phase, with smallholder farming becoming largely commercialized and new high-yielding technologies progressively adopted. Currently the smallholder cash crop sector is undergoing a phase of "extended structural change" as rural-urban migration reaches the point at which the absolute size of the agricultural labour force begins to decline (Morrison 1996).[22] Thus, many longhouse communities are losing people, especially young people, which makes it difficult to find the labour to maintain either subsistence or commercial agricultural pursuits.[23]

Contrary to Barlow (1997), however, the spread of smallholder cash crops was not due to diffusion from commercial estates (until the advent of oil palm). There were only ever five large rubber estates in Brooke-era Sarawak—three of them close to Kuching, one in the mid-Rejang and one at the far end of the state near Lawas (Kaur 1995)—and all were broken up in the post-war period. The more important influence was that of Chinese smallholders, who have a long history in rural areas of Sarawak; Chinese now make up 26 per cent of the Sarawak population, with a high proportion still involved in rural pursuits close to roads and towns (Cramb 1982).

The Spread of Chinese Smallholders into Shifting Cultivation Lands

Hakka Chinese had been settling in western Borneo since the middle of the 18th century, primarily mining gold within self-governing communities (*kongsi*) but also planting lowland rice and other crops (Chin 1981). In the first half of the 19th century increasing numbers crossed over from Sambas to Sarawak, settling around Bau and re-establishing the *kongsi* system. As well as mining antimony and gold, the Bau *kongsi* encouraged smallholders to produce food crops and livestock for local needs and cash crops such as pepper and gambier (*Uncaria gambir*), used in tanning and dyeing (Chin 1981, Chew 1990). In 1857, angered by newly imposed trade and other restrictions, the *kongsi* attempted unsuccessfully to overthrow the fledgling Brooke regime in Kuching. In retaliation, Charles Brooke led an Iban force to drive the insurgents back over the border, killing thousands in the process. Not surprisingly, Chinese mining and agricultural activities in Sarawak languished as a result.

Nevertheless, from the 1870s the numbers of Hakka and other Chinese groups in southwest Sarawak began to grow markedly, encouraged by new laws that made land available for intensive smallholder agriculture, particularly pepper and gambier.[24] Both contract or bonded labourers (*sinkeh*) and free settlers were recruited from China through Singapore to meet the perceived labour shortage. They leased small plots of land from the state on a "use-it-or-lose-it" basis or worked as tenants on the Borneo Company concession between Kuching and Bau (Chew 1990, Kaur 1995). The Second Rajah's view was that "anyone who takes the trouble to study the difference of cultivation between Dayaks and Chinese will easily arrive at the conclusion that one Chinese garden is of more value to the country than fifty Dayak holdings" (Criswell 1978: 139).

The spread of Chinese pepper and gambier planting in the Bau-Lundu area led to frequent conflict with Bidayuh shifting cultivators, whose fallow lands were being cleared and encroached upon. However, according to Chew, "Charles

Brooke, who wished to promote agricultural growth in the state and looked to the Chinese as the precursors of economic development, was not prepared to admonish the Hakka gardeners, even if it meant that Land Dayak [Bidayuh] customary law was being violated" (1990: 46). This official attitude was reflected in an 1875 order, intended to make ad hoc provision for the increasing number of Chinese farmers taking up land to the south and west of Kuching. The preamble to the order states, "it is common practice among the native communitys [sic] to make large clearings of old jungle, and afterwards abandon them." So the order allowed "squatters" (presumably Chinese smallholders without a title) to "occupy without interference land cleared and abandoned by others" (Porter 1967: 37). As Porter remarks, the order "suggests a curious misunderstanding on the part of Government, not simply of the practices permitted under native customary law but also of the biological demands the practices made on the land" (1967: 37).

As the Brooke state expanded its borders, Chinese smallholders were encouraged to settle in other districts. The most significant undertaking was an agreement in 1900 to settle Foochow colonists in the Rejang delta at Sibu, with a view to expanding rice production (Chin 1981, Chew 1990). More than 1,000 settlers arrived in 1901–2 and proceeded to plant wet rice. Early crop failures, disease, deaths and desertions gave way to food self-sufficiency by 1903, and by 1906 the settlers had turned to planting pepper and rubber. The first rubber trees were tapped in 1910, when the world price of rubber had risen dramatically. Rice cultivation was soon largely abandoned, and the Foochow capitalized on successive rubber booms to achieve economic prosperity. In Chew's words, "the Foochow settlers overcame the initial setbacks of padi crop failures, and in a spirit of independence, isolated themselves from other communities, catered to their own immediate needs of churches and schools, and relied upon rubber cultivation for their livelihood" (1990: 142).[25]

As in southwestern Sarawak, however, the notion that forest, whether primary or secondary, was "idle waste" became a serious problem as more settlers arrived and the Foochow farming settlements spread up the Rejang as far as Kanowit and downriver to Sarikei and Binatang. The preference of the Foochow for clearing secondary growth rather than primary forest, and their lack of any concept of farming that did not involve continuous cultivation, led them to encroach on Iban forest-fallow land, particularly with the rapid uptake of rubber planting after the 1910 boom. The resulting tensions peaked in 1925 with the outbreak of violence in the Binatang area. As Pringle remarks, "the Binatang incident of 1925 was only one spectacular symptom of a more basic problem, that of accommodating large numbers of Chinese farmers in a country of shifting cultivators" (1970: 13).

The spread of Hakka, Foochow and other Chinese pepper and rubber smallholders into Bidayuh and Iban lands thus prompted a second territorial strategy on the part of the Brooke state. As well as attempting to restrict the movement of shifting cultivators into primary forest through the creation of forest reserves, the state also moved to limit Chinese smallholder encroachment on the secondary forest of the shifting cultivators and, at the same time, to prevent the latter from "prematurely disposing of their land" to the Chinese. Consequently, the 1933 Land Rules introduced for the first time a distinction between native areas and mixed zones, with Chinese and other "non-native" farmers restricted to the latter (Porter 1967: 51). This racially based zoning system was carried over to the 1948 Land Classification Ordinance and incorporated in the 1958 Land Code. Despite a controversial attempt to overthrow it in the 1965 Land Bills, the land zoning remains in force (with important implications for the subsequent transition to plantation agriculture, discussed below).[26]

Figure 3.4 shows the distribution of mixed zone land, native area land, forest reserves and interior area land (a residual category) in the 1980s. As can be seen by comparing Figures 3.2 and 3.4, mixed zone land is found mainly

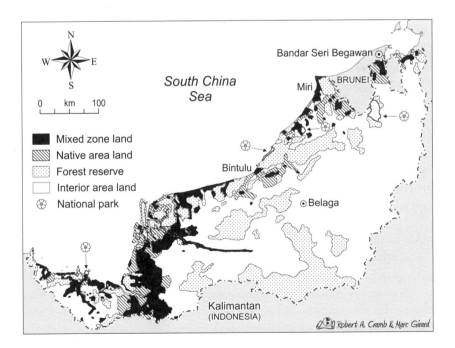

Figure 3.4 Sarawak. Land tenure classification

in regions of early Chinese smallholder expansion and skirts around most of the shifting cultivation lands of the Dayaks and the downriver settlements of the Malays and Melanau (except in the Saribas and Krian, where Iban farmers were more commercialized).

The Adoption of Rubber by Shifting Cultivators

As well as creating land conflicts, the juxtaposition of the intensive, commercial smallholdings of the Chinese and the more extensive, subsistence farming systems of the Dayaks and Malays exposed the latter to both Chinese agricultural practices and the trade networks that grew up to support them (T'ien 1953, Ward 1960).[27] This "demonstration effect" accelerated the adoption of cash crops (though Iban involvement in the 19th century trade in forest produce had prepared the way for new cash-earning activities; Cramb 1988a). Dayaks and Malays sometimes worked as labourers on Chinese pepper and rubber gardens and acquired planting material from them, as well as credit from traders in the local bazaars. Some experimented briefly with pepper and coffee planting in the 1890s, but it was the widespread adoption of rubber planting from around 1910 that transformed the rural economy. The Resident at Simanggang wrote in his memoirs: "I think that the years 1911–12 might be designated the Planting Era. Natives caught the rubber infection badly. Malays planted up all the land they could. Dyaks followed suit, and rubber banished all thoughts of tribal warfare and headhunting" (Ward 1966: 145).

Rubber fitted well into the Dayak agricultural cycle and grew well in a wide range of soil types. When it had been decided to establish a garden, the hill rice farms for that year would be cleared close to the longhouse or along a larger, more accessible stream, and the rubber seedlings would be inter-planted with the rice after weeding or in the stubble following the rice harvest. The forest regrowth would be slashed periodically in subsequent years to promote the growth of the young trees, and after 8–12 years the rubber would be ready for tapping (Bridges 1937: 66). This labour-economizing, "silvicultural" approach to crop establishment was already a familiar technique used in cultivating indigenous tree crops. In effect the rubber garden was a managed or enriched forest-fallow—one of Raintree and Warner's (1986) agroforestry pathways for the intensification of shifting cultivation.

In more commercially oriented regions such as Saribas District, many Iban landowners planted up to 10–20 hectares of rubber and employed Iban, Malay and Chinese labourers to tap their trees during the boom years (Cramb 2007). Communities in the more remote and less commercialized districts were slower to respond and planted much smaller areas, typically 1 or 2 hectares (Jensen

1966). By 1940, after three decades of smallholder rubber planting, there were 97,000 hectares under rubber, of which 92.5 per cent was in smallholdings (< 40 hectares), about equally divided between Chinese (mainly Hakka and Foochow) holdings (45 per cent by area) and "native" (mainly Iban and Malay) holdings (47.5 per cent). Chinese smallholdings averaged about 2.4 hectares, while smallholdings of the Iban and Malay averaged 0.6 hectares, notwithstanding the larger gardens of the Saribas Iban.[28] Sarawak's exports of rubber that year totalled 36,000 tons, valued at 26 million Straits dollars (Cramb 2007). In Reece's view, the predominance of smallholder rubber in Brooke Sarawak offered "an alternative form of economic ownership to that practiced in other parts of colonial Southeast Asia" (1988: 33).[29]

The post-war British colonial government built on this legacy with its Rubber Planting Scheme (RPS) to provide smallholders with improved, high-yielding planting material for both new planting and replanting, capitalizing on the 1950s rubber boom. The scheme was intended to wean hill farmers from their supposedly irrational attachment to shifting cultivation and establish them as viable commercial farmers, taking them into Barlow's "agricultural transformation" phase, though the technology of production has not kept pace with that in Peninsular Malaysia. Almost 100,000 hectares of smallholder rubber were planted in the first three decades of the RPS. Notwithstanding this effort, however, the deterioration in the world market for natural rubber led to a steady decline in production, from 51,000 tons worth RM122 million in 1960 to 20,000 tons worth RM15 million in 1972.

The RPS was suspended in 1972, and almost no further planting took place until its reintroduction in 1977, after which the area of high-yielding rubber continued to expand. Moreover, since 2000 the Sarawak government has initiated an expensive programme of rubber mini-estates, mainly on Iban fallow land, in which all the costs of establishing and managing the estates are met by the government, with net proceeds to be paid to the landholders. By 2005 the total area under rubber was stable at almost 157,000 hectares, of which 92 per cent was in smallholdings—now almost entirely owned by Dayaks and Malays rather than Chinese—and 8 per cent in the new mini-estates (Figure 3.1).[30] Just over 40 per cent of the smallholder rubber trees were high-yielding clones. In Dayak lands, rubber occupied between 25 per cent and 40 per cent of a longhouse territory (Cramb 2007).

However, the utilization of this extensive area of rubber has continued to decline. Rubber production reached a new low in 2001, when net exports amounted to just over 500 tons, worth only RM2.3 million. The downward trend in production reflected global market demand but also that Sarawak had entered Barlow's (1997) phase of "extended structural change", with fewer workers in the longhouses available and willing to tap their rubber trees. Even with the significant upsurge in rubber price since 2002, driven by demand from China, most of Sarawak's exports of rubber are in fact re-exports of processed rubber originating in Indonesian Borneo, where the opportunity cost of labour is lower.[31] Although the higher price has certainly helped farmers in hilly upland areas of Sarawak that are unsuited to oil palm (or were withheld from oil palm schemes), much of the tapping in areas such as the Saribas is now done by Indonesian workers, paid on a one-third share (*bagi tiga*) basis.[32]

The Rise of Dayak Pepper Cultivation

While the long-term involvement of shifting cultivators in rubber cultivation has been widely documented (Dove 1993, Wibawa *et al.* 2005), less often emphasized is the agricultural transformation brought about through the cultivation of pepper by Dayak smallholders, particularly since the 1970s (Wadley and Mertz 2005). This is partly because the highly intensive nature of pepper cultivation means that a holding of half a hectare is sufficient to fully employ (and fully support) a typical rural household. Thus, pepper may occupy only 1–2 per cent of an established longhouse territory while contributing perhaps 80 per cent of household income (Cramb 2007). This renders the transformation largely invisible to those, such as the chief minister, who want to see "modern agricultural development along the major trunk road with rows of plantations and villages well organized in centrally managed estates" (Cf. Figure 3.1).

Before the war only a small number of Dayaks had planted pepper, mostly in the vicinity of Chinese gardens. However, the post-war price boom for pepper was even more dramatic than that for rubber, prompting increasing numbers of Dayaks to take up pepper planting in the 1950s. The early Dayak gardens were small, usually around 50 vines. The largest would have been 100–200 vines, or less than 0.1 hectare. They were planted in hill farms following the rice harvest, and as they were unterraced and clean-weeded, the erosion was considerable—though the total area affected was not great. The gardens were fertilized with burnt earth and soil from under the longhouse. Later, organic fertilizers such as guano, prawn meal and soya meal were purchased. For insecticide, *tubai*—a traditional rotenone-containing root preparation of the climber *Derris elliptica*—was sprayed, using home-made bamboo pistons. Weed control was entirely by hand. The annual labour input for a 50-vine garden was probably around 25 man-days. The output would have been less than 100 kg, nevertheless returning enough in 1952 to buy almost twice the average household's annual rice requirement. Thus, the enterprise was kept to modest proportions in terms of land, labour and capital inputs and, despite being very remunerative, remained a small-scale activity, supplementary to hill rice cultivation (Best 1988, Cramb 1988a).

Pepper prices declined in the 1960s, but from 1968 there was a decade of steady price increases, encouraging a significant shift into pepper planting. A further development was the introduction of improved management practices based on a detailed programme of research that had been carried out by the Department of Agriculture since the colonial period. This involved the use of non-organic fertilizers and a range of measures to control the pests and diseases that had made pepper such a risky undertaking. The most serious disease, "footrot" or "sudden death" (caused by the soil fungus *Phytophthora cinnamomi*), though still without remedy, had nevertheless been thoroughly researched, and methods of containing it were now known. As a consequence, with adequate fertilization, pepper could be grown continuously on the same plot instead of requiring the plot to be periodically abandoned.[33]

The swing to pepper was given added impetus in 1972 with the introduction of the Pepper Subsidy Scheme (PSS). The subsidy was intended to cover 50 per cent of the establishment costs for a 200-vine garden, including fertilizers and hardwood supports for the pepper vines. The scheme met a genuine need among the Iban for capital. The Chinese shopkeepers would willingly provide fertilizer on credit within a year of the pepper harvest, on condition that the crop was then sold to them, but few would consider providing credit over a longer period. Institutional credit was also insufficient to meet the need. Notwithstanding the importance of the PSS, however, the majority of pepper vines were planted and maintained with the farmers' own resources (Cramb 1990a).

Thus, since the mid-1970s smallholder pepper cultivation has taken agricultural intensification in the shifting cultivation zone to a new level, in the sense of both a much higher input of labour and capital per unit of land as well as a total reliance on global markets and industrial inputs. Rather than pushing out the extensive margin, as with many boom crops, the growth of pepper production has involved a spatial concentration in small areas of existing territories, close to the longhouse and to transport routes (roads and rivers), consistent with both von Thunen and Boserup (Windle and Cramb 1997). In 2005 there were 64,100 households (more than half the total number of farm households in the state) cultivating nearly 12,700 hectares of pepper (0.2 hectare per household), producing more than 18,500 tons valued at more than RM115 million (Sarawak government 2007, Figure 3.1). Ninety per cent of these households were classified as "native"—almost all of them Dayaks—and these accounted for 88 per cent of the planted area.

So what had begun as a distinctly Chinese form of intensive agriculture was now predominantly a Dayak form of land use. Moreover, while rubber could be successfully adopted with a low level of technology (Barlow and Jayasuriya 1986), pepper has required the development of considerable technical, financial and managerial skills on the part of Dayak smallholders. At first pepper merely provided another cash-earning activity in a subsistence-oriented farming system still based on shifting cultivation. In the past two decades most Dayak farmers, even in quite remote areas, have become semi-commercialized or fully commercialized pepper farmers, operating intensive systems with considerable success (Cramb 1993, Wadley and Mertz 2005). Though highly dependent on the global market for pepper, they are not beholden to traders or moneylenders; in fact, many have substantial savings, as well as support from urban-based family members via remittances. With the increasing allocation of household labour to pepper production at the expense of subsistence shifting cultivation, the spatial concentration of production has been associated with significant disintensification in the surrounding secondary forest, in the sense that the frequency of cropping has fallen markedly and in some cases ceased altogether.

Intensification and Disintensification in Rice Farming

The post-war transition to smallholder cash crops, combined with an increase in rural-urban migration and the rising opportunity cost of agricultural labour, has had a major impact on Dayak and Malay rice production. From 1960 to 1985 rice output grew by an annual average of 3.2 per cent, faster than the rate of population growth (Cramb 1990a). However, this was entirely due to growth in the output of wet rice, which grew at 4.9 per cent; there was little or no change in the output, area cultivated or yield of hill rice over this period (Best 1988). The growth in wet rice production was due partly to a steady expansion of cultivated area and partly to a doubling in yield, from a low base of around 1.1 tons per hectare in the early 1960s to a still modest 2 tons per hectare in the mid-1980s.

The increase in yield was consistent with some improvements in land preparation and water control, and low levels of fertilizer application. There was evidence that the Assistance to Padi Planters Scheme (APPS), implemented by the Department of Agriculture since 1968, had been an important factor affecting both the area and the yield of wet rice, while the large-scale drainage and irrigation schemes scattered through the lowlands and accounting for 10 per cent of the wet rice area had contributed relatively little (Cramb 1992). So while there was a transition from traditional short-fallow swamp rice (*padi paya*) to bunded rainfed rice (*padi sawah*), cropped on an annual basis, there was no "green revolution" as such.

Following these years of steady growth, rice output halved from the mid-1980s to the late-1990s, before rising sharply to record levels in the mid-2000s (Figure 3.5). There were several factors at work here. First, while the yield of hill rice remained unchanged, the area cultivated began a downward trend of about 1.8 per cent per year that has continued to the present, from 85,000 hectares in 1985 to 65,000 hectares in 2006 (Figure 3.6). Second, the area of wet rice also fell, from 76,000 hectares in 1985 to a low of 55,000 hectares in 1997. Third, the yield of wet rice fell from around 2 tons per hectare in 1985 to 1.4 tons per hectare in 1998. All of these trends were consistent with a shift of labour effort away from rice production for subsistence towards commercial crops, especially pepper. Most of the rice farmers were also pepper farmers; hence, there was a sharp trade-off in labour use between the two crops as pepper cultivation expanded. In addition, the trends are consistent with the steady rural-urban migration that has been occurring since the mid-1980s, with longhouses losing population and labour as a result (Morrison 1996).

The gradual decline in hill rice cultivation is consistent with a trend throughout Southeast Asia (Padoch *et al.* 2007). In Sarawak it has involved farmers cultivating smaller annual clearings, cultivating intermittently rather than every year, and in many cases abandoning hill rice cultivation altogether (Cramb 2007). It is driven by many interrelated factors, including the conversion of fallow lands to oil palm estates and the loss of longhouse labour to

Figure 3.5 Sarawak. Rice production, 1985-2006

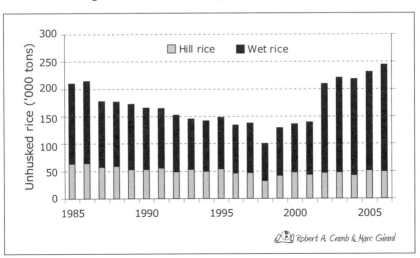

Figure 3.6 Sarawak. Area planted with hill rice, 1985-2006

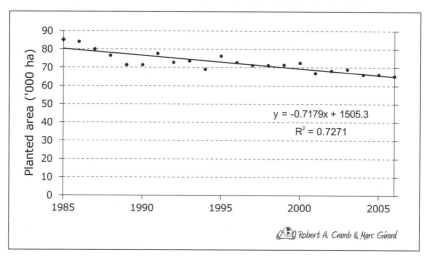

$$y = -0.7179x + 1505.3$$
$$R^2 = 0.7271$$

Robert A. Cramb & Marc Girard

off-farm pursuits, including education (Hansen and Mertz 2006). It can simultaneously involve more intensive rotations on smaller plots closer to roads and permanent farms (in some cases to supply urban markets with "traditional" foods), alongside the reversion of more remote fallow lands to old secondary forest (Hansen 2005, Hansen and Mertz 2006, Cramb 2007).

The decline in wet rice production, however, was spectacularly reversed from the late-1990s (Figure 3.5). While there has been some increase in area (12 per cent from 1997 to 2006), the most significant factor has been a doubling in yield from around 1.5 to 3 tons per hectare, particularly after 2002.[34] This is in response to the sharp increase in the world rice price since 2001. In addition, the federal government has been providing a fertilizer subsidy to wet rice farmers in Sarawak since 2000 under the "Skim Bantuan Baja Padi Persekutuan" (Federal Padi Fertilizer Assistance Scheme). More recently, added assistance has been given for land levelling, liming, pesticide and weedicide.[35] There has also been renewed investment in large-scale rice schemes.[36]

Thus, Dayak and Malay farmers in lowland areas have been investing in a significant intensification of wet rice cultivation—not by moving to multiple cropping but through land improvement and greater use of external inputs. In some cases, farmers residing in the uplands have also invested labour in lowland rice. Iban from the upper Saribas, having largely abandoned hill rice cultivation for pepper, have made use of the improved road network to "commute" to wet rice lands downriver that they have purchased, rented or borrowed, thus pursuing a form of spatial intensification as outlined by Pingali and Binswanger (1987).[37] Hence, the population pressure on these downriver sites has been greater than might appear from census data. As an example, of the 26 Iban households resident at Batu Lintang in the middle Saribas, 11 (42 per cent) had wet rice farms downriver near Betong in the 2000–1 season, while no one cultivated hill rice. At the same time, 24 households (92 per cent) worked on pepper gardens within their longhouse territory. No one tapped rubber, despite having extensive holdings, including a rubber mini-estate (Cramb 2007).

Smallholder Cash Crops and Land Tenure

The history of smallholder rubber and pepper not only demonstrates responsive and dynamic economic behaviour on the part of Dayak shifting cultivators but also shows that customary land tenure has not been an obstacle to the adoption and expansion of cash crops. Most of the developments described above have taken place on untitled land administered within longhouse communities according to their own *adat*. As indicated above, the customary tenure of Sarawak's shifting cultivators generally recognizes a permanent household cultivation right to land cleared from primary or old-growth forest. Consequently, individual households wanting to establish cash crops have been able to do so on their own land, without adversely affecting others in the community.

In the case of traditional Bidayuh land tenure, however, where the cultivation right circulated within a wider descent group, a decision by one member of the group to plant a crop such as rubber would unfairly remove a significant area of land from the common pool. Hence, in recent decades Bidayuh communities have almost universally moved by stages to divide the descent group's land, typically by first restricting the descent group to those with a common grandfather, then by dividing these lands among separate households. This corresponds to Boserup's analysis of the rising value of use rights with intensification, leading to greater individualization and permanency. However, it has not been a smooth process, often causing friction between close kin and creating opportunities for strategic behaviour, given the informal nature of the process and the limited capacity of village elders, let alone external authorities, to keep abreast of developments in what are now large and complex villages with many extra-village links and interests.

The Brooke state attempted to exercise some control over the planting of commercial crops on customary land, issuing planting grants, pepper garden permits and occupation tickets. These were issued without any supporting cadastral survey or map, so the state necessarily relied on the local knowledge of community elders for their enforcement. Nevertheless, they were held in high regard by Dayak smallholders, who treated them as government-sanctioned title to the plot in question, even when the rubber or pepper had been abandoned and the household had left the longhouse. In the post-war period there were no such titles issued for individual gardens; all that was required to be eligible for assistance from the RPS, PSS or APPS was the signature of the longhouse headman to certify that the land indeed belonged to the applicant.

Rather, the 1958 Land Code provided for the systematic survey of whole settlement areas and the issuing of individual titles to everyone with customary rights to land inside the area. From 1974 these titles became grants in perpetuity, so long as they remained with the original recipient or his/her heirs; once sold, they reverted to standard 99-year leases of state land, such as held by Chinese farmers in a mixed zone (Cramb 1982). However, the land titling process has been grindingly slow. By 1998, 40 years after the Land Code was introduced, only 88,000 hectares in the predominantly Iban Second Division had been titled, representing only 9 per cent of the total area and 11 per cent of customary land (Cramb 2007). So most customary land is still held by licence from the state. As mentioned, this has not been an obstacle for smallholder schemes, but it has become a major issue with the spread of oil palm plantations in the past 20 years.

As already discussed, Binswanger and McIntire (1987) hypothesized that intensification through cash crops would accelerate the individualization and transferability of specific use rights, giving land a market value and a value as collateral, resulting in increasing indebtedness, inequality, landlessness and tenancy. However, the spread of smallholder cash crops in Sarawak in the 20th century did not produce this outcome. Two interrelated factors have been important in this respect. First, the longhouse system of land tenure was, until recently, strong enough to resist the disequalizing effects of an unrestrained market in land. Second, the zoning of land since 1933 has meant that Dayak land could not be used as collateral for loans from Chinese moneylenders, so the latter could not hope to acquire land by foreclosing. Consequently, credit was restricted to short-term financing of inputs with the standing crop as collateral, particularly in the case of pepper. In the past ten years, with increased rural-urban migration, the erosion of community institutions, the spread of rural feeder roads, and especially the oil palm boom, there has been a complex and largely unrecorded process of land transactions involving extra-village actors (government officials, traders, professionals and others, many of them Dayaks), which may yet confirm the Binswanger-McIntire hypothesis.

The Transition from Smallholdings to Estates

The third agrarian transition in Sarawak has been by far the most rapid and pervasive, involving the expansion of plantation agriculture, particularly oil palm estates, throughout the territories previously occupied, utilized and claimed by shifting cultivation communities (Figures 3.1, 3.2 and 3.7). In fact, oil palm expansion has not only intensified land use in the shifting cultivation zone shown in Figure 3.2, but also, in a symbiotic relationship with logging, pushed the agricultural frontier into old-growth forests to an extent that was unimaginable 10 or 20 years ago, including both the peat swamp forests of the lowlands and the remaining hill forests of the deep interior. In the process, the long tradition of customary and smallholder land use and tenure that characterized Sarawak's agriculture up to the 1980s has been drastically modified or completely swept aside. Given the extensive areas involved and the desire to keep costs low, most of the labour for the private estate sector (as now in government land schemes) has been provided by legal and illegal Indonesian workers.

The two types of oil palm development, one involving the conversion of existing farmlands and secondary forest and the second the clearing of "primary"

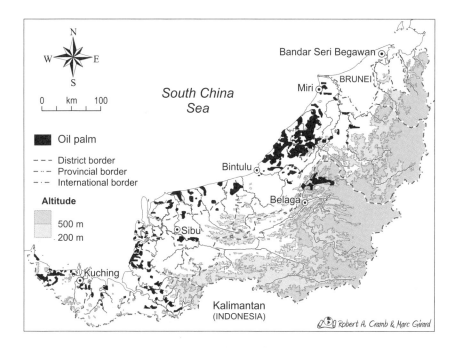

Figure 3.7 Sarawak. Distribution of oil palm, 2005

(logged-over) forest, have tended to be associated with different institutional modes, related to the pre-existing tenure status of the land. Where existing farm and fallow lands are involved and customary rights are recognized, oil palm has been planted largely in group or managed smallholdings and, more recently, in joint venture arrangements between landholders and plantation companies, as well as in independent smallholdings. Expansion into old-growth forest on what is deemed to be unencumbered state land has been the province of large-scale plantation companies, both public and private. However, even this distinction is blurred due to the contested nature of customary land rights in Sarawak, with areas of secondary forest and community forest reserves in the more recently settled central and northern regions of Sarawak being allocated by the government for private estate development on the presumption that no valid customary rights exist. A preliminary estimate, based on the aggregate planted area in the two types of oil palm development, is that around 80 per cent of the oil palm area has been converted from primary forest (following logging) and 20 per cent from secondary forest and other land uses.[38]

Oil Palm Estates on State Land

The first large-scale commercial planting of oil palm in Sarawak occurred in 1968 with the establishment of Sarawak Oil Palm (SOP) Bhd, a joint venture between the Sarawak government and the Commonwealth Development Corporation, which planted an area of just over 5,000 hectares of state land in northern Sarawak near Miri. This functioned from the outset as a commercial estate with largely Indonesian (Javanese) labour. From 1972, the Sarawak Land Development Board (SLDB) began to establish oil palm estates in the same region, primarily in forest reserves that had been designated for conversion to agriculture (Zain 1986). SLDB was established as the Sarawak equivalent of FELDA, which had been settling poor and landless Malay farmers in rubber and oil palm schemes in the Peninsula since 1956 (Sutton 1989, Fold 2000). The original policy was that the land schemes in Sarawak would also be subdivided following crop establishment, with titles being issued to individual settlers drawn from among the three major ethnic groups—Malays, Iban and Chinese. However, in 1974, due to the lack of interest from Malay and Chinese settlers and the government's unwillingness to make land settlement an exclusively Iban affair, it was decided to place a freeze on the recruitment of settlers. From that point the oil palm schemes were effectively run on conventional estate lines (Kedit 1974).

In 1981, when Abdul Taib Mahmud became chief minister, SLDB and SOP Bhd had the only oil palm estates in Sarawak, totalling about 20,000 hectares. However, this changed dramatically with the surge in profitability of oil palm, the closing of the agricultural frontier in Peninsular Malaysia, and the policies and legislative changes introduced by the Taib government (Cramb 2007). With cashed-up local timber companies such as WTK and Rimbunan Hijau, plantation companies from Peninsular Malaysia such as Perlis, Golden Hope, Boustead, and Tradewinds, and government-owned corporations such as FELDA Bhd keen to invest in large-scale oil palm estates, the Taib government streamlined access to state land for this purpose, including large areas subject to native customary rights claims. Speaking in 1982 at the launch of the Land Custody and Development Authority (LCDA), Taib's main vehicle for the transformation he had in mind, he "noted that many plantation-owning companies and individuals have found they cannot get enough suitable land in Peninsular Malaysia and are investing in agro-based projects in the Philippines and Papua New Guinea as a result. 'Why not come to Sarawak to invest? You can operate from Kuala Lumpur. Sarawak welcomes you,' Datuk Taib said" (Jong 1982). The two key mechanisms were the issuing of provisional leases (in effect, bypassing or deferring the question of customary land claims) and joint ventures with LCDA, which was "deemed to be a native" under the Land Code, thus permitting the private development of state and customary land outside a mixed zone.

Since the 1980s, with the Taib government's support, the SOP Group has grown into one of the largest private plantation companies, with more than 30,000 hectares of oil palm in Miri and Bintulu Divisions, generating revenue of RM85 million in 2005. SOP is now majority-owned by the Miri-based Shin Yang Group, headed by Datuk Ling Chiong Ho, which has extensive logging, plantation, shipping and construction interests and close connections with the political elite. LCDA also holds a substantial share. Shin Yang has been involved in controversial logging activities in the upper Baram as well as developing a large oil palm plantation in the forested lands in the upper Rejang vacated for the Bakun Dam project (Figure 3.7).

By 1980, notwithstanding its considerable assets, SLDB was making substantial losses and carried major liabilities. The Taib government moved by stages to privatize SLDB's operations and assets. In 1987 the management of SLDB was contracted to Sime Darby, a commercial plantation company, and it soon reported improved financial performance. In 1993 it was corporatized, and in 2000 its assets were transferred to Sarawak Plantation Bhd (SPB), a public company set up to facilitate a management buyout—though other players linked to Taib became involved, including the CEO of one of Sarawak's largest timber companies. By 2006 SPB had more than 24,000 hectares under oil palm in 13 estates, a legacy acquired from SLDB (Table 3.2). It was listed

Table 3.2 Sarawak. Oil palm-planted area by plantation category,
October 2006

Category	Planted area (ha)	% of total area
Independent smallholders	18,988	3.3
Organized smallholders		
– SALCRA	45,178	7.8
– FELCRA	26,980	4.6
Sub-total	72,158	12.4
Joint venture estates on customary land	33,193	5.7
Public-owned estates		
– Sarawak Plantation Bhd (formerly SLDB)	24,445	4.2
– FELDA Bhd	7,680	1.3
Sub-total	32,125	5.5
Private estates		
– Private-public joint ventures (LCDA)	65,527	11.3
– Private estates	359,049	61.8
Sub-total	424,576	73.1
Grand total	581,040	100.0

Sources: Sarawak Department of Agriculture; Ministry of Land Development,
Sarawak; Land Custody and Development Authority; Sarawak Plantation
Bhd.

on the Bursa Malaysia in August 2007, when the general manager announced
plans to secure 30,000–50,000 hectares of land in Indonesian Borneo for
further expansion of oil palm cultivation.

Apart from SOP and SLDB, the first new player in the post-1981
expansion of the estate sector was FELDA Bhd, operating as a public plantation
company. Perhaps hoping to replicate its large estate complex in eastern Sabah,
FELDA negotiated a provisional lease through LCDA to what was ostensibly
unencumbered state land near Lundu, in the southwestern corner of the
state. However, it encountered local resistance from Dayak claimants. This
was initially met with force but eventually resulted in a negotiated settlement.
Perhaps because of these difficulties, FELDA's operations have been confined
to this one estate of 7,680 hectares (Table 3.2).

Despite this early setback, from the 1990s the number of provisional
leases issued for oil palm development grew rapidly. Consequently, the area in
private and quasi-private (corporatized) oil palm estates increased from 20,300

hectares in 1990 to 456,700 hectares in 2006 and now represents almost 80 per cent of the total oil palm area (Table 3.2). In some cases the provisional lease was issued to a joint venture company formed between LCDA and a private investor, with LCDA holding 40 per cent of the equity. There were 20 such joint venture projects by 2006, with a total planted area of 65,500 hectares, mostly under oil palm. This accounted for 11 per cent of the total planted area in the state (Table 3.2). In most cases, however, provisional leases were issued directly to private plantation companies, including Sarawak-based timber companies and other local firms set up to exploit the oil palm boom (many with close political connections to the Taib government), as well as long-established plantation companies from Peninsular Malaysia. These purely private estates accounted for 359,000 hectares, or 62 per cent of the total oil palm area (Table 3.2). As well as the returns to investment in oil palm, developers and their associates often have the opportunity to profit up front from the clear-felling of timber in the lease area, thereby helping to finance the development (Fold and Hansen 2007). Another form of upfront benefit is obtained when a paper company formed for the purpose obtains a provisional lease and is then able to extract rent through its contractual relationship with a genuine developer.[39]

In theory, the provisional lease requires the lessee to identify any customary claims and negotiate acceptable arrangements (such as compensation) with the claimants before the lease can be confirmed. In practice, capital has been raised and land clearing commenced on the assumption that the provisional lease gives the company clear title to all the land falling within the perimeter of the lease area. Hence, longhouse communities who claimed customary rights to part or all of the land allocated for a private oil palm estate often knew nothing of the granting of a provisional lease until bulldozers arrived to clear the area for planting. When they protested they were mostly ignored, given notice to leave the area or, in the worst cases, subjected to violence. The response of a number of Dayak communities was to blockade estate access roads, impound bulldozers, and in many cases institute legal proceedings (IDEAL 1999, Cramb 2007). In two incidents in northern Sarawak, one in 1997 and the other in 1999, clashes between longhouse members and police or plantation company employees resulted in fatalities.

Such incidents represent the extreme outcomes of a widespread and ongoing conflict over land. In 2000 the Ministry of Land Development recorded 107 private oil palm estates, with a planted area of 263,000 hectares (an average of almost 2,500 hectares per estate). Of these, 24 were listed as having problems with their provisional lease. However, these 24 accounted for an approved area of 206,500 hectares and a planted area of 105,500 hectares (about 40 per cent of the private-sector-planted area at that time). Of the total approved area, 46,100 hectares (22 per cent) were under native customary rights claims and 16,400 hectares (8 per cent) were said to be occupied by "squatters". For individual estates the area under such land claims ranged from zero to 5,700 hectares in the case of SOP's Suai Estate, the latter comprising 87 per cent of the approved area.

Oil Palm Estates on Customary Land

Oil palm development on customary land was pioneered by the Sarawak Land Consolidation and Rehabilitation Authority (SALCRA), which was established in 1976 primarily to develop customary land "for the benefit of the owners". SALCRA's mode of operation has been to borrow public and donor funds on concessional terms for the capital costs of oil palm development. These costs are charged to the participants, who progressively pay back the amount and in time receive the net proceeds from the sale of their fruit. After taking adequate steps to "ascertain the wishes of the owners", SALCRA can declare a tract of land to be a "development area", thereby gaining the power to develop the land without affecting "the legal ownership of that land or any customary rights". In fact, the SALCRA Ordinance requires it to arrange a survey of land rights, and on completion of the development, landholders are issued with grants in perpetuity.

By 2006 SALCRA had established more than 45,000 hectares of oil palm, involving more than 12,500 participants. The finance for these schemes came from state and federal budgets and from development financiers such as the Asian Development Bank. While the labour was initially provided by the landholders, it now includes many Indonesian workers. Though the agency continues to be plagued by allegations of inefficiency and corruption, high palm oil prices and the passage of time have eased many of the participants' concerns. There is unmet demand from many upriver Iban communities who have requested a SALCRA scheme on their land. However, SALCRA's growth beyond southern Sarawak has been blocked by the government, with priority being given to joint venture projects (see below) in the central and northern regions.[40] The Federal Land Consolidation and Rehabilitation Authority (FELCRA) also undertakes in situ schemes for smallholders in Sarawak. It has been expanding rapidly in the past decade and had established nearly 27,000 hectares by 2006. Most of this is in downriver areas on titled Malay land. The area of land developed as group or managed smallholdings by these two agencies totalled about 72,000 hectares by 2006, or 12 per cent of the total oil palm area (Table 3.2).

Joint venture estates on customary land are those established under the Taib government's "Konsep Baru" (New Concept) policy, launched in 1995 (Ngidang 2002). Under this policy the customary landholders agree to assign their land rights to the LCDA, which then forms a joint venture company with a private-sector partner.[41] A consolidated land title covering 5,000 hectares or more is issued to the joint venture company for a 60-year period (two cycles of oil palm). Following a rough ground survey of individual holdings within the lease area, the private investor pays the value of the land to the owners, pegged at RM1,200 per hectare. Of this, 10 per cent is paid up front in cash, 30 per cent is invested in a government unit trust scheme, and 60 per cent is regarded as the landowners' (30 per cent) equity in the company. The private-sector partner has 60 per cent equity and LCDA 10 per cent. Landholders receive no title to their land but can expect to receive dividends according to the area of land contributed to the project. In effect, then, the private plantation company leases the land from LCDA, acting as trustee for the landholders, in return for 60 per cent of the profits, and it can then manage the estate as if it was a private concern.[42] Landholders can obtain employment on the estate but are not involved in any management decisions or financing arrangements; in practice, most of the labour, particularly for harvesting, is now provided by Indonesian workers.

By 2006 there were 34 joint venture projects on customary land, with a total of 33,000 hectares planted with oil palm (Table 3.2). Thus, only 6 per cent of the total oil palm area is in Konsep Baru projects, less than half the area under group or managed smallholdings. Over a third of the area in joint venture schemes is in one large project, the Kanowit Oil Palm Project, commenced in 1995 with Boustead Holdings Bhd as the joint venture partner. However, the project has given rise to a number of issues (Ngidang 2002, 2005; IDEAL 1999; Matsubara 2003). Some of the early concerns expressed by both participants and non-participants were the following: there was little opportunity to pursue alternative land-use options as almost all the land was planted with oil palm; the workers received low wages and disliked plantation work; there was no involvement in the management of the joint venture company; there was uncertainty about the level of future dividends; there was uncertainty about the status of the land at the end of the project, or if the project failed. Similar concerns were expressed by customary landholders in other areas earmarked for Konsep Baru schemes (Songan and Sindang 2000, Ngidang 2000, Majid-Cooke 2006). These concerns seemed well justified in that by 2007, more than a decade after the policy was launched, none of the joint venture companies had issued dividends. In 2008 this led to protests and blockades in the Kanowit project (Thien 2008), calling into question the future of the joint venture approach.[43]

Independent smallholders have grown at a rapid rate in the past decade, particularly in the vicinity of the large-scale estates and palm oil mills in northern Sarawak (Majid-Cooke *et al.* 2006). In this case, customary landowners are developing their land using their own labour and capital, without assistance from a government agency. However, some who have good connections with Chinese traders obtain finance to establish their plantations, and in some cases wage labour is employed, including Indonesian workers who prefer this arrangement to working on private or government estates. Some of this planting is clearly pre-emptive, that is, to prove to government agencies that may seek to allocate the land for a private estate or joint venture estate that it is being productively used (Majid-Cooke 2006). Subsidized or supported smallholders are those who plant on individual lots, perhaps in a contiguous area, with varying degrees of support from government agencies such as the Department of Agriculture or the Malaysian Palm Oil Board. By 2006 there were around 3,418 smallholders (independent and supported), with an average planted area of 5.3 hectares. Though the area planted by these smallholders was only 3 per cent of the total in 2006 (Table 3.2), the growth has been rapid, from only 670 hectares in 1990 to the current figure of 19,000 hectares—more than doubling in the past six years. Subsidized or supported smallholdings account for only 2,040 hectares, or 12 per cent of the total area of smallholders in these two categories. Thus, the growth in smallholdings has occurred largely without government support, and despite some opposition from sections of the government favouring estate development.

It is true that, unlike rubber or pepper, oil palm cultivation displays economies of scale in first-stage processing, and the harvested product is not storable, so there is a need for a minimum planted area within a maximum distance from a mill for timely and efficient conversion of fresh fruit bunches into palm oil. This is the technical basis of the argument for large-scale, centrally managed production systems. However, private estates are only one such system. The varied history of oil palm expansion in Sarawak outlined above shows that a spectrum of institutional arrangements, including those that rely on smallholder initiative, can deliver these benefits. In particular, the adoption of oil palm by unassisted smallholders in those regions with access to mills is an unsurprising extension of Sarawak's long history of autonomous smallholder development.

Conclusion

Taking the long view of agrarian transition in Sarawak gives a new perspective on the processes of agricultural intensification and expansion. Successive waves of agricultural expansion, overlapping in time and space, have entailed the intensification of previous land and forest use, as people, crops and farming systems have spread through the landscape. These transitions have entailed not only agro-ecological or land-use change but also the (sometimes violent and often contested) reworking of property rights, territorial control and modes of agricultural organization—from nomadic bands and longhouse villages to independent family smallholdings as well as large private and public estates. The result has been a distinctive layering or imbrication of socio-ecological landscapes such that elements of pre-existing landscapes influence both the possibilities and constraints of subsequent transitions, seen most clearly in the political geography of recent oil palm expansion.

The first and second agrarian transitions have been driven not only by demographic change, as highlighted in Boserup's initial contribution, but also by sociocultural shifts, improvements in transportation, market trends and fluctuations, new crop technologies, and changes in agrarian laws and policies, as later writers in the Boserupian tradition (as well as Boserup herself) have elaborated. Broadly consistent with the Boserup hypothesis, the intensification of land use has resulted in the long-term, aggregate-level growth of food output and incomes in Sarawak, beyond the growth in rural population, though obviously not all members of the population have shared equally in this growth. Food output grew due to the areal expansion of hill rice cultivation and both the areal expansion and increased yield of wet rice cultivation. Incomes grew due to the complementary incorporation of cash crops, notably rubber, pepper and oil palm, in previously subsistence farming systems. This growth in income, though assisted by government programmes in the post-war era, has been largely due to innovation and investment on the part of both independent and village-based smallholders.

The environmental constraints to intensification in Sarawak have, however, been more significant than anticipated by Boserup, such that further intensification of hill rice cultivation has not been seen as a productive use of household labour. Rather, in recent decades, the spatial concentration of smallholder agricultural production in more favourable locations (both environmentally and in relation to infrastructure) has resulted in parallel trends of intensification (and commercialization) in these locations and disintensification in the more remote forest-fallow lands. Hence, the widespread and remarkably persistent belief that shifting cultivation would result in a downward spiral of land degradation and deforestation has not been borne out.

Boserup's theory also depicts an evolution of land tenure from rotating use rights of relatively low exchange value to permanent, individual rights of high exchange value as land use intensifies. Binswanger and McIntire's extension of this theory predicts the emergence of indebtedness, inequality, landlessness and tenancy as commercial tree crops are incorporated in the land-use system. In Sarawak, however, this evolutionary pathway has not been as inexorable or unidirectional as the theory suggests. Most shifting cultivators have in fact recognized permanent household rights to cleared land within the community territory since the pioneer phase. In some cases land scarcity has subsequently induced communities to adopt rotating use rights to maintain equitable access to land for subsistence, thus reversing Boserup's hypothesized sequence. Moreover, the recent intensification and commercialization of smallholder agriculture has not led to significant landlessness and tenancy, as seen in many parts of Indonesia and the Philippines. Among the reasons are Sarawak's low population pressure and relative land abundance (particularly for intensive crops such as pepper), the widespread provision of capital inputs by the state in the form of planting grants, legal safeguards against debt-induced loss of customary land rights, and the high rate of permanent rural-urban migration associated with Malaysia's rapid economic growth and structural transformation. However, as communities become "less bounded socially and spatially",[44] informal land transactions are occurring on an increasing scale, particularly in the now extensive peri-urban zone, warranting further detailed research to trace the processes involved and their implications.

The third agrarian transition, involving plantation agriculture in conjunction with commercial logging, is beyond the scope of Boserup's analysis, being driven not by local population growth or smallholder market opportunities but a radical policy shift favouring the expansion of private investment from both inside and outside Sarawak, coupled with the large-scale importation of Indonesian labour (the latter a demographic consequence rather than a cause of intensification). No doubt the output of commodities—both timber and palm oil—from the remaining forestlands has increased substantially as a result of this intensification of resource use. However, unlike earlier land-use changes, the broad-scale impacts on the hill and swamp forest environments have been devastating, while the benefits to local livelihoods have been, at best, marginal (as in many SALCRA schemes) and, at worst, totally detrimental (as in failed joint venture schemes or private estates that exclude customary landholders).

As it turns out, then, Charles Brooke's desire to see "the jungle falling left and right" has been more than fulfilled in recent decades, but his agrarian populist vision of "people settled over what are now lonely wastes" has not been realized. In fact, the notion of "lonely wastes" has taken on a new meaning in a landscape of logged-over and heavily degraded forests, end-to-end oil palm plantations, and half-deserted longhouses (only partly offset by the presence, not of settler homesteads as Brooke envisaged, but the hidden-away barracks of migrant Indonesian labourers). Similarly, Taib Mahmud's vision of "modern agricultural development along the major trunk road with rows of plantations" has been largely achieved, but his long-held plans to have "villages well organized in centrally managed estates with a stake of their own in them" has proved more contentious and elusive. Expansionist agrarian policies that fail to acknowledge the complex historical and geographical layering of customary usages and claims—that treat the landscape, in other words, as if it was "levelled terrain on which to build (dis)utopias" (Scott 1998: 89)—will invariably encounter the kinds of frustration, losses, conflict and resistance experienced during the third agrarian transition in Sarawak.

Notes

1. I am grateful to Rodolphe De Koninck, Wolfram Dressler, Tania Li, John McCarthy, Ole Mertz and Jonathan Rigg for helpful comments on a previous draft. The research for this paper was funded by an Australian Research Council Discovery Grant. Parts of this chapter draw on Cramb (2007), with further updating of data and analysis.

2. Diary entry for 21 February 1867, "Extract from the Diary of Charles Brooke", cited in Pringle (1970: 304n).

3. Quoted in *Sarawak Tribune*, 9 Sept. 1984.

4. As background, it can be noted that Sarawak was a province of the Brunei Sultanate before 1841, the private colony of the Brooke Rajahs from 1841 to 1941, occupied by the Japanese military from 1941 to 1945, and a British Crown Colony from 1946 to 1963. It became a state in the Federation of Malaysia in 1963, though retaining control over land and forest resources.

5. I am using "imbricated" here following Fernandes (2006), who uses the word to describe the overlapping and mutually structuring nature of state and non-state legal institutions, as illustrated by the traditional landholding institutions (*comunidades*) of Goa. As well as institutional layering, however, I am using the term to include the layering of associated land-use systems.

6. Much as Hoskins (1955) depicted the deep historical layering that gave rise to the English landscape.

7. In her 1965 essay Boserup leaves aside the question of the causes of population pressure, focusing only on the impacts. In a later paper she allows for recursiveness between food production technology and population growth but argues that "demographic trends in primitive populations are influenced not only by food technology but also by health, transport, and war technologies, and by the system of organization, which could be called 'administrative technology'" (Boserup 1976: 22).

8. Boserup (1975) subsequently expanded her subsistence framework to allow for differences in such factors as rural infrastructure, opportunities to produce a surplus for trade, and labour migration.

9. Note De Koninck's earlier contributions in French on this theme, e.g., "Enjeux et stratégies spatiales de l'État en Malaysia", *Hérodote* 22 (1981): 84–115; "La

paysannerie comme fer de lance territorial de l'État: le cas de la Malaysia", *Cahiers des Sciences Humaines* 22, no. 3–4 (1986): 355–70.

10. Evidence of pre-Austronesian habitation in Borneo goes back 40,000 years, though these original inhabitants probably remained in coastal environments and did not penetrate far into the interior rainforest. The Austronesians came from southern China via Taiwan and the Philippines, probably arriving in northern Borneo by about 2500 BC (Bellwood 1992, 2007; Bellwood *et al.* 1995). Though they brought with them rice, millet and other crops, it seems likely that rice cultivation was initially confined to scattered lowland environments, while those who moved into the interior developed a largely hunter-gatherer adaptation. Hence, Hunt and Rushworth's (2005) evidence for rice cultivation in the Niah catchment in the first millennium BC or earlier is not necessarily inconsistent with the suggestion of Doherty *et al.* (2000) that rice farming was not widespread in Sarawak until medieval times.

11. McKinley (1978) and Helliwell (1992) argue that Bornean hill farmers such as the Iban may have adopted hill rice cultivation rather than invest in lowland agriculture as much for political reasons as demographic, moving into the uplands to escape domination by coastal elites. McKinley points out that Malay outposts of the Melaka Sultanate were being established at river mouths around Borneo at the same time as the Iban migrations out of the Kapuas began.

12. Conklin (1957) introduced the distinction between "pioneer" and "established" shifting cultivation, the former involving the expansion of shifting cultivation through the felling of primary forest and the latter the practice of rotational shifting cultivation in secondary forest. However, this distinction is also blurred in the Sarawak context; historically, the pioneer phase typically overlapped with the established phase from the outset, and the two coexisted for long periods, perhaps centuries, such that both primary and secondary forest were cleared until only islands (*pulau*) of primary forest remained as reserves and a full forest-fallow cycle was established throughout a given territory (Freeman 1970, Cramb 2007).

13. A possible exception is the homeland of the Kenyah at the headwaters of the Iwan tributary of the Kayan River, in what is now East Kalimantan. This is said to be grassland of *Imperata cylindrica* degraded due to the earlier pressure of a growing population and the adoption of shifting cultivation. Once conflict with the neighbouring Kayan had ceased, successive groups of Kenyah moved out of the upper Iwan into new territory and abandoned their place of origin (Chin 1985). This is not true of the Iban and Bidayuh, representatives of whom still occupy their ancestral lands in West Kalimantan.

14. As with the Iban, "during a period of expansion, the Kayan either displaced or subjugated the populations of the areas they entered" (Ngo 2003: 172). The Kenyah expansion followed that of the Kayan, beginning in earnest around 1900 and continuing today. Guerreiro and Sellato (1984) provide an insightful analysis of the traditional process of migration, based on a contemporary account.

15. These included the Kanowit, Kejaman, Sekapan, La'anan, Punan Bah, Berawan and others, each now occupying no more than two or three longhouses interspersed among Iban, Kayan and Kenyah settlements. In the 20th century the Iban reached the Limbang Valley, partly displacing another group, the Bisaya, who cultivate mainly

wet rice with the aid of buffaloes and are more closely related to Sabahan ethnic groups to the north (Blandoi 2004). The analysis of reasons for the limited spread of the buffalo (*Bubalis bubalis*) into Sarawak is another story.

16. The current chief minister of Sarawak, a Melanau, is alleged to have commented on the eve of taking up his post in 1981: "Now we will get the land back from the Iban!" His subsequent policies lend some credence to this alleged remark, presumably alluding to a sense of historical dispossession.

17. See Janowski (2004) for an important account of the development and intensification of rice growing in the Kelabit highlands, a somewhat different story from that described here for other Dayak areas.

18. Bugis and Hakka Chinese smallholders also developed coconut holdings in this area, since renamed Asajaya.

19. Displaced hunter-gatherer groups such as the Bukitan moved just ahead of the Iban frontier into new territory and progressively adopted Iban agricultural practices and ways of life, becoming "Ibanized" in the process. A survey in 1978 in the Anap River in Bintulu Division included a single Bukitan longhouse at the headwaters of a stream occupied by Iban settlements (Cramb and Dian 1979b). Many of the residents had intermarried with local Iban, and both the Iban and Bukitan languages were spoken in the longhouse. There was also an elderly woman of the related Lugat tribe, of which no separate communities remained, presumably having been previously absorbed into Bukitan society. The Bukitan practised shifting cultivation of rice but were less proficient than their Iban neighbours, perhaps because of their greater reliance on hunting and gathering. This has also been observed among the "semi-settled" Penan of the upper Rejang (Brosius 1986, State Planning Unit 1987).

20. However, Chin (1985) found the Kenyah system at Long Selatong in the Baram to be similar to that of the Bidayuh, with land circulating within descent groups. The Selakau studied by Schneider (1978) also followed both forms. This variation comes down to the critical choice of whether to divide a household's accumulated landholdings at the time of the formation of an offshoot household or to keep the land as an undivided pool for all the descendants of the first feller to draw on. In both situations the head of the original household (in Iban the *bilek asal*) retains authority (*kuasa*) over the allocation of land. In the case of fruit and other valuable forest trees, it seems that among all Dayak groups these are typically not divided on household partition but remain the property of the descent group, again subject to the control of the senior household (Rousseau 1974, Whittier 1978, Sather 1990, Cramb 2007).

21. Pedersen *et al.* (2006) found that in the Model Forest Management Area south of Bintulu, Iban shifting cultivation and commercial logging were able to coexist, because of a clear demarcation between the secondary forest used by the shifting cultivators and the primary forest that was within the logging concession, but they acknowledge this might have been a special case.

22. Despite the difficulties in adequately defining and measuring the agricultural labour force and the rural population in Sarawak, it is likely that by the 1980s or 1990s Sarawak had reached what Tomich *et al.* (1995: 12–7) refer to as the "structural

transformation turning point", when the absolute size of the agricultural labour force peaks and begins to decline. This is leaving aside the Indonesian agricultural workforce brought in to work on the oil palm plantations.

23. A study of two Iban longhouses in Saribas District found that from 1980 to 2001, the resident population had remained steady in one and declined by 37 per cent in the other, consistent with census data showing an overall decline of the Iban population in this largely rural district (Cramb 2007: 302). Correspondingly, in 2001 only 8 per cent of households in the two longhouses cultivated hill rice and none tapped rubber trees; instead, 96 per cent concentrated on small plots of pepper.

24. An 1876 proclamation offered grants of land for 99 years at a nominal rent to Chinese pepper and gambier planters, who were offered assistance to migrate to Sarawak for the purpose (Porter 1967: 38–9). These grants were initially made in the vicinity of Kuching. In 1896 the provisions of the Land Regulations were applied to the issue of grants for pepper and gambier throughout Sarawak. A permit from the Resident was required to open or extend a garden. However, "squatters" cultivating pepper and gambier with "care and diligence" were allowed to continue.

25. This, coupled with their strategic position in relation to the Rejang's vast timber resources, provided the basis for their economic domination and political influence in post-war Sarawak (Leigh 1988).

26. The 1962 Land Committee argued in support of Brooke and British colonial land policy that "though we appreciate the urgent need of the Chinese for more land … we believe the native must be prevented from disposing of his land until he has been better educated in how to use it properly" (Government Printing Office 1962). Nevertheless, informal long-term leasing arrangements outside mixed zones, particularly between Chinese farmers and Iban and Bidayuh landholders, persist to the present day.

27. The Hakka and Foochow dialect groups, as well as dominating in the agricultural sector, were the dominant groups in first-stage trading in the rural bazaars (T'ien 1953).

28. However, many Iban households had several separate "holdings".

29. Though John McCarthy (pers. comm.) makes the point that in areas of Sumatra away from the plantation belt a similar pattern of smallholding prevailed.

30. A correction was made to estimated rubber area from 1995 to allow for senile gardens and those cleared for other uses. The revised estimate for this year (172,000 hectares) probably represents the peak expansion of smallholder rubber. (Dominic Dares, Sarawak Department of Agriculture, pers. comm., 15 Dec. 2008.)

31. Dominic Dares, Sarawak Department of Agriculture, pers. comm., 15 Dec. 2008.

32. Joseph Blandoi, pers. comm., 7 Aug. 2008.

33. Reid (1995) argues that traditional forms of pepper cultivation caused extensive deforestation in insular Southeast Asia in the centuries after 1400.

34. This also coincides with improved data collection techniques by the Department of Agriculture, meaning that earlier yields may have been underestimated.

35. Dominic Dares, pers. comm., 15 Dec. 2008.

36. Ole Mertz, pers. comm., 16 Mar. 2009. However, yields from these schemes are not significantly higher than other wet rice areas.

37. The Saribas Iban adopted this strategy in the 1960s and early 1970s, when rubber and pepper prices were low and hill rice yields were poor, but at that time had to reside permanently downriver during the rice season.
38. This is supported by Hansen (2005), who, in a study of land-use change between 1972 and 2002 in the Niah catchment in northern Sarawak (at the heart of the oil palm zone), found that oil palm increased from 0 per cent to 41 per cent of the catchment area during this period and that about 88 per cent of the oil palm area in 2002 had been converted from logged-over forest and 12 per cent from former shifting cultivation lands.
39. The payment (premium) for a provisional lease is RM741 per hectare, whereas the "market" rate for such land is around ten times that amount.
40. This block had been partly lifted by 2008, though only for the remote Kapit Division, which was unattractive to commercial investors, having no road access.
41. LCDA was established in 1982. As in the case of SALCRA, LCDA is deemed to be a native, which gives it power to deal in native customary land. It acts as an intermediary rather than a land development agency as such.
42. Fold and Hansen (2007) also argue that the Konsep Baru policy was merely a way to alienate more land for private development at a time when most of the suitable state land was becoming scarce.
43. By August 2008, in response to the protest, a small "interim dividend" had been paid in one of the six Kanowit estates and "advances" were paid or promised for the others.
44. Tania Li's phrase.

References

Angelsen, A., *Forest Cover Change in Space and Time: Combining the von Thunen and Forest Transition Theories*, World Bank Policy Research Working Paper 4117. Washington, D.C.: World Bank, 2007.

Barlow, C., "Growth, Structural Change and Plantation Tree Crops: The Case of Rubber", *World Development* 25 (1997): 1589–607.

Barlow, C. and S. Jayasuriya, "Stages of Development in Smallholder Tree Crop Agriculture", *Development and Change* 17 (1986): 635–58.

Barlowe, R., *Land Resource Economics: The Economics of Real Estate*, 4th edition. New Jersey: Prentice Hall, 1986.

Barton, H. and T. Denham, "Vegeculture and Social Life in Island Southeast Asia", in *Why Cultivate? Anthropological and Archaeological Approaches to Foraging-Farming Transitions in Southeast Asia*, ed. G. Barker and M. Janowski. Leiden: KITLV Press, in press.

Bauer, P.T., *The Rubber Industry: A Study in Competition and Monopoly*. London: Longman Green, 1948.

Bellwood, P., "The Prehistory of Borneo", *Borneo Research Bulletin* 24 (1992): 7–15.

_____, *Prehistory of the Indo-Malaysian Archipelago*, 3rd edition. Canberra: ANU E Press, 2007.

Bellwood, P., J.J. Fox and D. Tryon, "The Austronesians in History: Common Origins and Diverse Transformations", in *The Austronesians: Historical and Comparative Perspectives*, ed. P. Bellwood, J.J. Fox and D. Tryon. Canberra: Department of Anthropology, Australian National University, 1995, pp. 1–16.

Best, J.R., "Change over Time in a Farming System Based on Shifting Cultivation of Hill Rice in Sarawak, Malaysia", *Agricultural Administration and Extension* 29 (1988): 69–84.

Binswanger, H. and J. McIntire, "Behavioural and Material Determinants of Production Relations in Land-Abundant Tropical Agriculture", *Economic Development and Cultural Change* 36 (1987): 73–99.

Black, J.D., "The Extensive vs. the Intensive Margin", *Journal of Farm Economics* 11 (1929): 331–3.

Blaikie, P. and H. Brookfield, *Land Degradation and Society*. London and New York: Methuen, 1987.

Blandoi, J., *Bisaya Journey*. Kuching: Lee Ming Press, 2004.

Blaug, M., *Economic Theory in Retrospect*, 5th edition. New York: Cambridge University Press, 1997.

Boserup, E., *The Conditions of Agricultural Progress: The Economics of Agrarian Change under Population Pressure*. Chicago: Aldine, 1965.

_____, "The Impact of Population Growth on Agricultural Output", *Quarterly Journal of Economics* 89 (1975): 257–70.

_____, "Environment, Population, and Technology in Primitive Societies", *Population and Development* 2 (1976): 21–36.

_____, *Population and Technology*. Oxford: Basil Blackwell, 1981.

Bridges, W.F.N., "A Report on Rubber Regulation in Sarawak". Kuching, 1937.

Brookfield, H.C., "Intensification and Disintensification in Pacific Agriculture: A Theoretical Approach", *Pacific Viewpoint* 13 (1972): 30–48.

_____, "Intensification Revisited", *Pacific Viewpoint* 25 (1984): 15–44.

Brosius, J.P., "River, Forest and Mountain: The Penan Gang Landscape", *Sarawak Museum Journal* 36 (1986): 173–84.

_____, "A Separate Reality: Comments on Hoffman's 'The Punan: Hunters and Gatherers of Borneo'", *Borneo Research Bulletin* 20 (1988): 81–105.

Chew, D., *Chinese Pioneers on the Sarawak Frontier, 1841–1941*. Singapore: Oxford University Press, 1990.

Chin, J.M., *The Sarawak Chinese*. Kuala Lumpur: Oxford University Press, 1981.

Chin, S.C., "Shifting Cultivation: A Need for Greater Understanding", *Sarawak Museum Journal* 25 (1977): 107–28.

_____, *Agriculture and Resource Utilisation in a Lowland Rainforest Kenyah Community*, Special Monograph No. 4. Kuching: Sarawak Museum, 1985.

_____, "Shifting Cultivation and Logging in Sarawak", in *Logging against the Natives of Sarawak*. Petaling Jaya: Institute for Social Analysis, 1989, pp. 57–64.

Chisholm, M., *Rural Settlement and Land Use: An Essay in Location*, 3rd edition. London: Hutchinson, 1979.

Cleary, M. and P. Eaton, *Borneo: Change and Development*. Singapore: Oxford University Press, 1992.

Colchester, M., *Pirates, Squatters and Poachers: The Political Ecology of Dispossession of the Native Peoples of Sarawak*. London: Survival International; Petaling Jaya: Institute for Social Analysis, 1989.

Conklin, H.C., *Hanunoo Agriculture: A Report on an Integral System of Shifting Cultivation in the Philippines*. Rome: Food and Agriculture Organization, 1957.

Cramb, R.A., "A Survey of Chinese Farmers in the Sungai Tengah Area". Farm Management Report No. 1, Department of Agriculture, Sarawak, 1982.

———, "The Commercialisation of Iban Agriculture", in *Development in Sarawak: Historical and Contemporary Perspectives*, ed. R.A. Cramb and R.H.W. Reece. Melbourne: Centre of Southeast Asian Studies, Monash University, 1988a, pp. 105–34.

———, "Shifting Cultivation and Resource Degradation in Sarawak: Perceptions and Policies", *Review of Indonesian and Malaysian Affairs* 22 (1988b): 115–49.

———, "Explaining Variations in Bornean Land Tenure: The Iban Case", *Ethnology* 28 (1989): 277–300.

———, "The Role of Smallholder Agriculture in the Development of Sarawak: 1963–88", in *Socio-Economic Development in Sarawak: Policies and Strategies for the 1990s*, ed. Abdul Majid Mat Salleh, Hatta Solhee and Mohd. Yusof Kasim. Kuching: AZAM, 1990a, pp. 83–110.

———, "Forestry and Land Use in Sarawak: Reply to John Palmer", *Borneo Research Bulletin* 22 (1990b): 44–6.

———, "Problems of State-Sponsored Land Schemes for Small Farmers: The Case of Sarawak, Malaysia", *Pacific Viewpoint* 33, no. 1 (1992): 58–78.

———, "Shifting Cultivation and Sustainable Agriculture in East Malaysia: A Longitudinal Case Study", *Agricultural Systems* 42 (1993): 209–26.

———, "Farmers' Strategies for Managing Acid Upland Soils in Southeast Asia: An Evolutionary Perspective", *Agriculture, Ecosystems and Environment* 106 (2005): 69–87.

———, *Land and Longhouse: Agrarian Transformation in the Uplands of Sarawak*. Copenhagen: NIAS Press, 2007.

Cramb, R.A. and Z. Culasero, "Landcare and Livelihoods: The Promotion and Adoption of Conservation Farming Systems in the Philippine Uplands", *International Journal of Agricultural Sustainability* 1 (2003): 141–54.

Cramb, R.A. and J. Dian, "A Social and Economic Survey of the Dalat Extension Region". Planning Division, Department of Agriculture, Sarawak, 1979a.

———, "A Social and Economic Survey of the Bintulu Extension Region". Planning Division, Department of Agriculture, Sarawak, 1979b.

Cramb, R.A. and I.R. Wills, "The Role of Traditional Institutions in Rural Development: Community-Based Land Tenure and Government Land Policy in Sarawak, Malaysia", *World Development* 18 (1990): 347–60.

Criswell, C.N., *Rajah Charles Brooke: Monarch of All He Surveyed*. Kuala Lumpur: Oxford University Press, 1978.

De Koninck, R., "The Peasantry as the Territorial Spearhead of the State: The Case of Vietnam", *Sojourn: Social Issues in Southeast Asia* 11 (1996): 231–58.

———, "The Theory and Practice of Frontier Development: Vietnam's Contribution", *Asia Pacific Viewpoint* 41 (2000): 7–21.

_____, "Challenges of the Agrarian Transition in Southeast Asia", *Labour, Capital and Society* 37 (2004): 285–8.

De Koninck, R. and W.D. McTaggart, "Land Settlement Processes in Southeast Asia: Historical Foundations, Discontinuities, and Problems", *Asian Profile* 15 (1987): 341–56.

Doherty, C., P. Beavitt and E. Kerui, "Recent Observations of Rice Temper in Pottery from Niah and Other Sites in Sarawak", *Bulletin of the Indo-Pacific Prehistory Association* 20 (2000): 147–52.

Dove, M.R., "Smallholder Rubber and Swidden Agriculture in Borneo: A Sustainable Adaptation to the Ecology and Economy of the Tropical Forest", *Economic Botany* 47 (1993): 136–47.

Ellen, R., "The Distribution of *Metroxylon sagu* and the Historical Diffusion of a Complex Traditional Technology", in *Smallholders and Stockbreeders: Histories of Foodcrop and Livestock Farming in Southeast Asia*, ed. P. Boomgaard and D. Henley. Leiden: KITLV Press, 2004, pp. 69–106.

Elson, R.E., *The End of the Peasantry in Southeast Asia: A Social and Economic History of Peasant Livelihood, 1800–1900s*. London: Macmillan, 1997.

Fernandes, J., "Interlegality and the Mutual Imbrication of State and Community: A Study of the *Comunidades* of Goa". Paper presented at annual meeting of the Law and Society Association, Baltimore, 6 July 2006.

Fold, N., "Oiling the Palms: Restructuring of Settlement Schemes in Malaysia and the New International Trade Regulations", *World Development* 28 (2000): 473–86.

Fold, N. and T.S. Hansen, "Oil Palm Expansion in Sarawak: Lessons Learned by a Latecomer?" in *Environment, Development and Change in Rural Asia-Pacific: Between Local and Global*, ed. J. Connell and E. Waddell. London and New York: Routledge, 2007, pp. 146–66.

Freeman, J.D., *Report on the Iban*. London: Athlone, 1970.

Geddes, W.R., *The Land Dayaks of Sarawak*. London: HMSO, 1954a.

_____, "Land Tenure of Land Dayaks", *Sarawak Museum Journal* 6 (1954b): 40–51.

Gibbons, D.S., R. De Koninck and I. Hasan, *Agricultural Modernization, Poverty, and Inequality: The Distributional Impact of the Green Revolution in Regions of Malaysia and Indonesia*. Aldershot: Gower, 1980.

Government Printing Office, "Report of the Land Committee". Kuching: 1962, p. 14.

Guerreiro, A.J. and B.J.L. Sellato, "Traditional Migration in Borneo: The Kenyah Case", *Borneo Research Bulletin* 16 (1984): 12–28.

Hansen, T.S., "Spatio-Temporal Aspects of Land Use and Land Cover Changes in the Niah Catchment, Sarawak, Malaysia", *Singapore Journal of Tropical Geography* 26 (2005): 170–90.

Hansen, T.S. and O. Mertz, "Extinction or Adaptation? Three Decades of Change in Shifting Cultivation in Sarawak", *Land Degradation and Development* 17 (2006): 135–48.

Harrisson, T. *The Malays of South-West Sarawak before Malaysia: A Socio-Ecological Survey*. London: Macmillan, 1970.

Hart, G., A. Turton, B. White, B. Fegan and T.G. Lim, eds., *Agrarian Transformations: Local Processes and the State in Southeast Asia*. Berkeley: University of California Press, 1989.

Hatch, T., *Shifting Cultivation in Sarawak: A Review*, Technical Paper No. 8. Kuching: Soils Division, Department of Agriculture, Sarawak, 1982.

Helliwell, C., "Evolution and Ethnicity: A Note on Rice Cultivation Practices in Borneo", in *The Heritage of Traditional Agriculture among the Western Austronesians*, ed. J.J. Fox. Canberra: Department of Anthropology, Research School of Pacific Studies, Australian National University, 1992, pp. 7–20.

Heppell, M., L. Melaka and E. Usen, *Iban Art: Sexual Selection and Severed Heads*. Leiden: Art Books; Amsterdam: Kit Publishers, 2005.

Hong, E., *Natives of Sarawak: Survival in Borneo's Vanishing Forests*. Penang: Institut Masyarakat, 1987.

Hoskins, W.G., *The Making of the English Landscape*. London: Hodder and Stoughton, 1955.

Hunt, C.O. and G. Rushworth, "Cultivation and Human Impact at 6000 cal yr B.P. in Tropical Lowland Forest at Niah, Sarawak, Malaysian Borneo", *Quaternary Research* 64 (2005): 460–8.

IDEAL, *Tanah Pengidup Kitai: Our Land Is Our Livelihood*. Sibu: Integrated Development for Eco-Friendly and Appropriate Lifestyle, 1999.

_____, *Not Development but Theft: The Testimony of Penan Communities in Sarawak*. Sibu: Institute for Development and Alternative Lifestyle, 2000.

INSAN, *Logging against the Natives of Sarawak*. Petaling Jaya: Institute for Social Analysis, 1989.

Janowski, M., "The Wet and the Dry: The Development of Rice Growing in the Kelabit Highlands, Sarawak", in *Smallholders and Stockbreeders: Histories of Foodcrop and Livestock Farming in Southeast Asia*, ed. P. Boomgaard and D. Henley. Leiden: KITLV Press, 2004, pp. 139–62.

Jensen, E. "Money for Rice: The Introduction of Settled Agriculture Based on Cash Crops among the Ibans of Sarawak, Malaysia". Unpublished report, Danish Board for Technical Cooperation with Developing Countries, 1966.

Johnson, C.L., "Government Intervention in the Muda Irrigation Scheme: 'Actors', Expectations and Outcomes", *Geographical Journal* 166 (2000): 192–214.

Jong, K.C., "A Year Focused on an Overwhelming Problem", *Borneo Bulletin* (10 Apr. 1982): 11.

Kaur, A., "The Babbling Brookes: Economic Change in Sarawak, 1841–1941", *Modern Asian Studies* 29 (1995): 65–109.

Kedit, P.M., "From Subsistence Farmer to Estate Worker", *Sarawak Gazette* 100 (1974): 25–32.

King, V.T., *The Peoples of Borneo*. Oxford: Blackwell, 1993.

Langub, J., "Some Aspects of Life of the Penan". Paper presented at Orang Ulu Cultural Heritage Seminar, Miri, 21–23 June 1988.

_____, "Native Customary Rights: Indigenous Perspectives". Paper presented at Malaysian Forest Dialogue, Kuala Lumpur, 22–23 Oct. 2007.

Le Gros Clarke, C.D., *The Blue Report*. Kuching: Sarawak government, 1935.

Leigh, M. "The Spread of Foochow Commercial Power before the New Economic Policy", in *Development in Sarawak: Historical and Contemporary Perspectives*, Monash Papers on Southeast Asia No. 17, ed. R.A. Cramb and R.H.W. Reece. Melbourne: Centre of Southeast Asian Studies, Monash University, 1988, pp. 179–90.

Lembat, G. "Native Customary Land and the *Adat*". Paper presented at the Seminar on NCR Land Development, Santubong Resort, Sarawak, 1994.

Li, T.M., "Local Histories, Global Markets: Cocoa and Class in Upland Sulawesi", *Development and Change* 33 (2002): 415–37.

Majid-Cooke, F., "Expanding State Spaces Using 'Idle' Native Customary Land in Sarawak", in *State, Communities and Forests in Contemporary Borneo*, ed. F. Majid-Cooke. Canberra: ANU E Press, 2006, pp. 25–44.

Majid-Cooke, F., D. Ngidang and N. Selamat, "Learning by Doing: Social Transformation of Small-holder Oil Palm Economies of Sabah and Sarawak, Malaysia". Report to UNESCO by Universiti Malaysia Sabah, Universiti Malaysia Sarawak and Universiti Sains Malaysia, 2006.

Masyhuri and R.A. Cramb, "A Socio-Economic Assessment of Land-Use Practices in a Transmigration Settlement on Acid Soils in South Kalimantan, Indonesia", in *Plant-Soil Interactions at Low pH: Principles and Management*, ed. R.A. Date, N.J. Grundon, G.E. Rayment and M.E. Probert. Dordrecht: Kluwer, 1995, pp. 685–7.

Matsubara, T., "Society and the Land: Contemporary Iban Society, Development Policy, and the Value of Native Customary Rights Land in Sarawak, Malaysia". MSocSc thesis, Institute of East Asian Studies, Universiti Malaysia Sarawak, 2003.

McKinley, R., "Pioneer Expansion, Assimilation, and the Foundations of Ethnic Unity among the Iban", *Sarawak Museum Journal* 26 (1978): 15–27.

Morris, H.S., "A Problem in Land Tenure", in *The Societies of Borneo: Explorations in the Theory of Cognatic Social Structure*, ed. G.N. Appell. Washington, D.C.: American Anthropological Association, 1976, pp. 110–20.

———, "The Coastal Melanau", in *Essays on Borneo Societies*, Hull Monographs on South-East Asia No. 7, ed. V.T. King. Oxford: Oxford University Press, 1978, pp. 37–58.

———, *The Oya Melanau*. Kuching: Malaysian Historical Society (Kuching Branch), 1991.

Morrison, P.S., "Urbanization and Rural Depopulation in Sarawak", *Borneo Research Bulletin* 27 (1996): 127–37.

Ngidang, D., "Native Customary Land Rights, Public Policy, Land Reform and Plantation Development in Sarawak", *Borneo Review* 8 (1997): 63–80.

———, "People, Land and Development: Iban Culture at the Crossroads", in *Iban Culture and Development in the New Reality*, ed. D. Ngidang, S.E. Sanggin and R.M. Saleh. Kuching: Dayak Cultural Foundation, 2000, pp. 34–49.

———, "Contradictions in Land Development Schemes: The Case of Joint Ventures in Sarawak, Malaysia", *Asia Pacific Viewpoint* 43 (2002): 157–80.

———, "Transformation of the Iban Land Use System in Post Independence Sarawak", *Borneo Research Bulletin* 34 (2003): 62–78.

———, "Deconstruction and Reconstruction of Native Customary Land Tenure in Sarawak", *Southeast Asian Studies* 43 (2005): 47–75.

Ngo, T.H.G.M., "A New Perspective on Property Rights: Examples from the Kayan of Kalimantan", in *Borneo in Transition: People, Forests, Conservation, and Development*, ed. C. Padoch and N.L. Peluso. Kuala Lumpur: Oxford University Press, 2003, pp. 170–86.

Nicolaisen, I., "Pride and Progress: Kajang Response to Economic Change", *Sarawak Museum Journal* 36 (1986): 75–116.

Padoch, C., "Land Use in New and Old Areas of Iban Settlement", *Borneo Research Bulletin* 14 (1982a): 3–14.

_____, *Migration and Its Alternatives among the Iban of Sarawak*. The Hague: Martinus Nijhoff, 1982b.

Padoch, C., E. Harwell and A. Susanto, "Swidden, Sawah, and In-between: Agricultural Transformation in Borneo", *Human Ecology* 26 (1998): 3–20.

Padoch, C. and N.L. Peluso, eds., *Borneo in Transition: People, Forests, Conservation, and Development*, 2nd edition. Kuala Lumpur: Oxford University Press, 2003.

Padoch, C. *et al.*, "The Demise of Swidden in Southeast Asia? Local Realities and Regional Ambiguities", *Danish Journal of Geography* 107 (2007): 29–41.

Pedersen, M., O. Mertz and G. Hummelmose, "The Potential for Coexistence between Shifting Cultivation and Commercial Logging in Sarawak", in *State, Communities and Forests in Contemporary Borneo*, ed. F. Majid-Cooke. Canberra: ANU E Press, 2006, pp. 181–93.

Peluso, N.L., "Whose Woods Are These? Counter-Mapping Forest Territories in Kalimantan, Indonesia", *Antipode* 27 (1995): 383–406.

Peluso, N.L. and P. Vandergeest, "Genealogies of the Political Forest and Customary Rights in Indonesia, Malaysia, and Thailand", *Journal of Asian Studies* 60 (2001): 761–812.

Pelzer, K., *Pioneer Settlement in the Asiatic Tropics*. New York: American Geographical Society, 1948.

Pingali, P.L. and H.P. Binswanger, "Population Density and Agricultural Intensification: A Study of the Evolution of Technologies in Tropical Agriculture", in *Population Growth and Economic Development: Issues and Evidence*, ed. D.G. Johnson and R.D. Lee. Madison: University of Wisconsin Press, 1987, pp. 27–56.

Porter, A.F., *Land Administration in Sarawak*. Kuching: Sarawak Government Printer, 1967.

Pringle, R., *Rajahs and Rebels: The Ibans of Sarawak under Brooke Rule, 1841–1941*. Ithaca: Cornell University Press, 1970.

Raintree, J.B. and K. Warner, "Agroforestry Pathways for the Intensification of Shifting Cultivation", *Agroforestry Systems* 4 (1986): 39–54.

Reece, R.H.W., "Economic Development under the Brookes", in *Development in Sarawak: Historical and Contemporary Perspectives*, Monash Papers on Southeast Asia No. 17, ed. R.A. Cramb and R.H.W. Reece. Melbourne: Centre of Southeast Asian Studies, Monash University, 1988, pp. 21–34.

Reid, A., "Humans and Forests in Pre-Colonial Southeast Asia", *Environment and History* 1 (1995): 93–110.

Rousseau, J., "The Social Organisation of the Baluy Kayan". PhD thesis, Cambridge University, 1974.

_____, "Kayan Agriculture", *Sarawak Museum Journal* 25 (1977): 129–56.

_____, "The Kayan", in *Essays on Borneo Societies*, Hull Monographs on South-East Asia No. 7, ed. V.T. King. Oxford: Oxford University Press, 1978, pp. 78–91.

_____, "Kayan Land Tenure", *Borneo Research Bulletin* 19 (1987): 47–55.

Ruthenberg, H., *Farming Systems in the Tropics*, 3rd edition. Oxford: Clarendon Press, 1980.

Sandin, B., "The Westward Migrations of the Sea Dayaks", *Sarawak Museum Journal* 7 (1956): 57–81.

_____, *The Sea Dayaks of Borneo before White Rajah Rule*. London: Macmillan, 1967.

Sarawak government, *Agricultural Statistics of Sarawak 2005*. Kuching: Department of Agriculture, Sarawak, 2007.

Sather, C., "Trees and Tree Tenure in Paku Iban Society: The Management of Secondary Forest Resources in a Long-Established Iban Community", *Borneo Review* 1 (1990): 16–40.

Schneider, W.M., "The Selako Dayak", in *Essays on Borneo Societies*, Hull Monographs on South-East Asia No. 7, ed. V.T. King. Oxford: Oxford University Press, 1978, pp. 59–77.

Scott, J.C., *Weapons of the Weak: Everyday Forms of Peasant Resistance*. New Haven: Yale University Press, 1985.

_____, *Seeing Like a State: How Certain Schemes to Improve the Human Condition Have Failed*. New Haven: Yale University Press, 1998.

Seavoy, R.E., "The Transition to Continuous Rice Cultivation in Kalimantan", *Annals of the Association of American Geographers* 65 (1973): 218–25.

Sellato, B., *Nomads of the Borneo Rainforest: The Economics, Politics, and Ideology of Settling Down*. Honolulu: University of Hawaii Press, 1994.

Songan, P. and A. Sindang, "Identifying the Problems in the Implementation of the New Concept of Native Customary Rights Land Development Projects in Sarawak through Action Research". Paper presented at "Borneo 2000: Environment, Conservation and Land", Sixth Biennial Conference of Borneo Research Council, Kuching, 10–14 July 2000.

State Planning Unit, "Report on the Effects of Logging Activities on the Penans in Baram and Limbang Districts". Submitted to the Sarawak State Cabinet Committee on Penan Affairs, 1987.

Strickland, S.S., "Long-Term Development of Kejaman Subsistence: An Ecological Study", *Sarawak Museum Journal* 36 (1986): 117–72.

Sutton, K., "Malaysia's FELDA Land Settlement Model in Time and Space", *Geoforum* 20 (1989): 339–54.

Thien, T., "Angry NCR Landowners Act against Company", *Malaysiakini*, http://www.malaysiakini.com/print.php?c=news&id=84914, accessed 23 June 2008.

T'ien, J.K., *The Chinese of Sarawak: A Study of Social Structure*. London: London School of Economics and Political Science, 1953.

Tiffen, M. and M. Mortimore, "Malthus Controverted: The Role of Capital and Technology in Growth and Environment Recovery in Kenya", *World Development* 22 (1994): 997–1010.

Tomich, T.P., P. Kilby and B.F. Johnston, *Transforming Agrarian Economies: Opportunities Seized, Opportunities Missed*. Ithaca: Cornell University Press, 1995.

Vandergeest, P. and N.L. Peluso, "Territorialization and State Power in Thailand", *Theory and Society* 24 (1995): 385–426.

Van Kooten, G.C., *Land Resource Economics and Sustainable Development: Economic Policies and the Common Good*. Vancouver: UBC Press, 1993.

Vincent, J.R. and R. Mohamed Ali, *Environment and Development in a Resource-Rich Economy: Malaysia under the New Economic Policy*. Cambridge: Harvard University Press, 1997.

Wadley, R. and O. Mertz, "Pepper in a Time of Crisis: Smallholder Buffering Strategies in Sarawak, Malaysia, and West Kalimantan, Indonesia", *Agricultural Systems* 85 (2005): 289–305.

Ward, A.B., *Rajah's Servant*, Southeast Asia Program Data Paper No. 61. Ithaca: Cornell University, 1966.

Ward, B.E., "Cash or Credit Crops? An Examination of Some Implications of Peasant Commercial Production with Special Reference to the Multiplicity of Traders and Middlemen", *Economic Development and Cultural Change* 8 (1960): 148–63.

Whittier, H.L., "The Kenyah", in *Essays on Borneo Societies*, Hull Monographs on South-East Asia No. 7, ed. V.T. King. Oxford: Oxford University Press, 1978, pp. 92–122.

Wibawa, G., S. Hendratno and M. van Noordwijk, "Permanent Smallholder Rubber Agroforestry Systems in Sumatra, Indonesia", in *Slash-and-Burn Agriculture: The Search for Alternatives*, ed. C.A. Palm, S.A. Vosti, P.A. Sanchez and P.J. Ericksen. New York: Columbia University Press, 2005, pp. 222–32.

Windle, J. and R.A. Cramb, "Remoteness, Roads and Rural Development: Economic Impacts of Rural Roads in Sarawak, Malaysia", *Asia Pacific Viewpoint* 38 (1997): 37–53.

WRM/SAM, *The Battle for Sarawak's Forests*. Penang: World Rainforest Movement and Sahabat Alam Malaysia, 1989.

Zain, A., "Experience of SLDB in Developing State and Native Customary Land: A Case Study", in *Kemajuan Menerusi Pembangunan Tanah*, Report of the Fourth Development Seminar, Bintulu, Apr. 1986. Kuching: Angkatan Zaman Mansang, pp. 95–136.

4

Claiming Territories, Defending Livelihoods: The Struggle of Iban Communities in Sarawak

Jean-François Bissonnette
University of Toronto[1]

Introduction

During the second half of the 20th century, the Malaysian state of Sarawak experienced fast and far-reaching environmental change. Forested areas that until then had remained largely untouched were extensively logged, reshaping livelihoods and local economies (Brookfield 1990, King 1993, Drummond and Taylor 1997). In addition, logging activities cleared the way for the establishment of oil palm plantations. As Cramb explains (Cf. Chapter 3), the private oil palm boom that started in the 1980s constitutes a distinct phase in the contemporary agrarian history of Sarawak. Most oil palm cultivation in Sarawak nowadays is carried out by the private sector, which relies on a corporate plantation model applied to state land. As a result, the pressure exerted on land resources by private oil palm plantations has caused numerous disputes pertaining to landownership between native communities and oil palm corporations. This is especially true in Miri Division, where oil palm plantations have been expanding at a rapid pace, becoming the main feature of the landscape, often encircling the territories of native communities.

This chapter describes some of the consequences of the pressure exerted by corporate oil palm companies as they try and convert into plantations territories claimed by native communities as customary land. It analyzes some of the practices and actions of two Iban communities to assert their land rights and pursue economic development despite threats of land dispossession from oil palm corporations. In doing so, it pays attention to the economic, legal and institutional contexts in which land struggles take place. The empirical data provided in this study is based on field observations and interviews carried out during the months of May to September 2006, mainly in Miri Division, located in the northeast of Sarawak (Figure 4.1). The research also draws on the study of official documents. The field observations offer insights into forms of political power deployed at the local scale to legitimize native territorial rights (Kerkvliet 2005; Li 2000, 2002a, 2002b). Overall, the study contributes to the analysis of the complex issue of native land and resource entitlement in Sarawak.

In the first section, recent rural transformations are summarized in order to explain the context of land disputes in the countryside of Sarawak. The history of Iban territoriality formation is also reviewed, so as to provide some background to the current landownership and access issues. Recent regulations undermining native customary rights to land are equally referred to. The second section provides a description of the two communities studied, with a focus on the strategies used to defend land rights and cope with the changing institutional context of land rights. This leads to an analysis of the implications of the actions and practices deployed by these communities to legitimize their land rights. The communities at the core of the study are at the forefront of dramatic transformations; they provide examples that may help understand the impacts of institutional and economic transformations on rural populations in East Malaysia.

The Context of the Problem

Why Territories Are at Stake

In Sarawak, sizeable tracts of land have been converted into oil palm plantations, mostly since the mid-1990s, with the result that by 2005 oil palm cultivation area in Sarawak surpassed 500,000 hectares. In addition, more than 400,000 hectares have been transformed into or allocated for forestry plantations.[2] By September 2006, 271,000 hectares[3] of native customary land, divided up into 41 schemes,[4] had been alienated for oil palm plantation expansion as part of the rural development strategy of public-private joint ventures, referred to as "Konsep Baru"[5] (Ministry of Land Development 2006). Although only a small proportion of these schemes have materialized (Cf. Chapter 3), these figures reveal more the government's intention than its capacity to convince native communities of the benefits of large-scale oil palm cultivation schemes.

The Konsep Baru in Sarawak forms one of the pillars of a broader pan-Malaysian modernization ideology, "Vision 2020", which sets national socio-economic development objectives to be reached by the year 2020. In Sarawak, this policy is epitomized by a number of major development and infrastructure projects, such as the Bakun Dam (Cf. Chapter 3). With the rise in palm oil prices on international markets, the state has deliberately facilitated the expansion of oil palm plantations in Sarawak. This has conferred an unprecedented merchant value to large areas of so-called empty or idle state land. Large tracts of cultivable land legally unencumbered with customary rights have already been allocated to private plantation companies. As a result, customary territories already alienated or claimed by native communities— approximately 1.5 million hectares—are now targeted for further oil palm estate expansion.

According to official statements, most land suitable for agriculture and not already allocated to other purposes should be transformed into commercial plantations. Oil palm estate expansion is legitimized by a discourse emphasizing the imperative of modernization and socio-economic development in Sarawak. Oil palm plantation expansion as proposed by the state would benefit many "impoverished and backward native communities" as well as the entire state in an era requiring effective "competition" on the global stage.[6] But the growing demand for land suitable for oil palm cultivation generates confrontations between many native communities and oil palm corporations.

The state's capacity to equitably deliver improvement of living conditions is increasingly questioned in Sarawak. The reputation of the developmental state was tarnished in the postcolonial era by alleged collusion and favouritism practices in the granting of logging concessions (Osman 2000: 984; Colchester 1993; Majid-Cooke 1997). More recently, the same nepotistic practices were reported in provisional land leases granted by the state (Ngidang 2005: 67). The relative depletion of natural forest resources for timber extraction in the early 1990s in Sarawak created a void in the state's finances and, according to many, threatened to weaken patronage linkages. This would explain in part, according to some, the oil palm expansion phase that began in Sarawak in the 1990s. Oil palm expansion finds its place within complex governance not only at the state scale, but also at the local scale. When oil palm corporations threaten to encroach on native land, they interfere with pre-existing customary territorial organization.

The legal definition of native customary land and its interpretation are critical to the settlement of overlapping claims between native communities and oil palm corporations. Nevertheless, conditions imposed by the state on Iban communities for accessing formal land titles, such as clear individual landownership, are often incompatible with existing communal landownership in many native communities. In this context, many Iban communities, backed by activists and detractors of governmental interventions, are keen to criticize the state's development agenda. Native customary land titles granted during the colonial era, although lawful, nowadays are often interpreted by governmental land management authorities as mere recognition of land utilization rights over a communal territory. They do not constitute land titles—in the modern legal sense of the term—and therefore native land rights are sometimes overlooked when the state issues provisional leases for private plantation development. As a result, in many rural constituencies native customary land rights became one of the main issues that shaped the 2006 state election results (Puyok 2006). In 2007, between 100 and 150 communities filed court cases against the state itself, its land regulation agencies (such as the Sarawak Land and Survey Department) and private oil palm/forestry companies.[7] Given the stakes of these decisions, and despite numerous verdicts rendered in favour of native communities, all cases were still pending at different court levels in July 2007. The same situation prevailed in January 2009, according to the Borneo Project, an NGO specializing in indigenous rights advocacy active in Sarawak.[8]

As the state of Sarawak adopts deregulation policies to facilitate capital investments and economic growth through oil palm plantation expansion, it also enforces constraining regulations pertaining to native land access and ownership. Therefore, commoditization of native territories and of agricultural production appears as the ultimate goal of the state. With regulations aimed at commoditizing more and more areas of state land and native customary land, the actions of the state intertwine with the globalization process as they both contribute to increasing the level of capitalization in agriculture. In this context, native territories become the locus of conflicts and the site in which power struggles over land rights come into play.

The Production of Native Customary Land

Native communities now find themselves in a regime of rapid oil palm plantation expansion. The native customary land that is often at stake in the regime of plantation expansion has changed throughout Sarawak history. It is therefore useful to revisit the historical evolution of native customary land in order to understand land disputes between native communities and oil palm corporations. Although Iban notions of territoriality existed before Western colonization, native customary land was shaped by the colonial administrations that had a prominent role in the overall Iban community's territorial evolution. After Sarawak was handed over by the Sultanate of Brunei to British interests, two main entities successively governed it: the three Rajahs of the Brooke dynasty (from 1841 to 1941) and the British colonial administration (from 1946 to 1963). Then, in 1963, Sarawak joined the Malaysian Federation, but following a historical agreement negotiated with the federal government, it retained power over a number of jurisdictions, such as land and natural resources. The Brooke colonial dynasty established a regime of indirect rule over the vast hinterland of Sarawak. The three White Rajahs of Sarawak applied widely different types of intervention, but the three concurred to force the settlement of Iban populations. Before colonial intervention, Iban populations had been engaged in a rapid territorial expansion through the practice of pioneer colonization (Ngidang 2005: 53).

The colonial state intended to circumscribe the area utilized by every Iban community, replacing moving frontiers with stable ones. Although the effective capacity of the state to control natural resource utilization was fairly limited outside a range of a few kilometres from local administration centres, especially under James Brooke (1841–68), basic land regulations progressively reached most areas of Sarawak (Pringle 1970). With the consolidation of colonial power and the formation of modern state institutions, pioneer colonization became supervised by the administration's agencies. This occurred under Vyner Brooke (1917–46) and even more under direct British colonial rule, which began in 1946 (Reinhardt 1970). These measures favoured the concept of fixed community, represented in the power structure by an appointed ethnic leader, drawing its legitimacy from ties with the colonial administration (Sutlive 1988: 144). State territorial control was further asserted by the cooptation of community political organizations.

Driven by a paternalistic approach that required subservience to its rule, the Brooke colonial administration was nevertheless concerned with not disrupting the native "way of life". Consequently, from 1863 onwards it instituted native customary rights, a progressive recognition of a set of institutions governed by traditional laws—the Dayak *adat*—which was embedded in the colonial state legislation. Iban institutions were thereby gradually integrated into and subordinated to Sarawak's state institutions (Cramb and Wills 1990, Pringle 1970). From then on, the native territorial realm being demarcated, the colonial state went on to distribute the remaining lands to non-natives, mainly ethnic Chinese people. This system was eventually crystallized in the ethnic-based land classification introduced alongside the implementation of the Torrens System in 1933, for instance, formalizing interdictions for natives to sell their land to non-natives.

The Chinese, whether new or long-established migrants, constituted the bulk of the labour force in the market economy, whose development was encouraged by the successive Brooke administrations. Chinese labourers and peasants provided the tax revenue for the expenses of the government. Until the widespread adoption of rubber cultivation in the early 20th century, native communities were largely left out of the market economy. Yet involvement in the market economy by specific native groups had long existed through commercial activities such as sago cultivation by the Melanau communities of Mukah Division. Nevertheless, under the Brooke regime the state of Sarawak did not encourage large-scale crop plantation expansion, unlike the North Borneo Chartered Company (Sabah) or the Dutch government of the East Indies (Indonesia). The aim of the administration was apparently mostly humanistic, although its *raison d'être* was ultimately to assert power over a territory and its inhabitants.

The means of the colonial administration being relatively limited, Iban customary land, the *pemakai menoa*, could not be fully surveyed. Neither were individual land titles requested by most Iban households, which did not see how they would benefit from them as they already considered themselves owners of the land they were cultivating. For instance, the Land Order of 1933 led to the creation of native land reserves, "which were not to be subdivided and for which individual title would not be issued" (Ngidang 2005: 58). In fact, the native customary rights guaranteed the continuity of community-based traditional institutions. According to the traditional individual land tenure system, every household is entitled to usufruct rights over land plots used for cultivating paddy through shifting cultivation. This basic traditional activity is complemented by hunting and gathering on fallow lands. The customary natural resource governance model also comprises communal units, such as primary forest areas, waterways, water catchments and longhouse land (Cramb 1986). In a memorandum issued in 1939, colonial officials recognized the said model along with its wider sociocultural and religious implications (Bulan 2006: 47).

The result of customary land recognition is that communal and individual plots, even today, cannot be mortgaged and used as collateral. This has two consequences. On the one hand, the native peasantry lack access to bank loans, which hinders further investments in commercial agriculture. On the other hand, this has historically prevented indebtedness and seizure of lands, protecting native populations from agrarian capitalism and widespread landlessness (Doolittle 2003).

With generalized adoption of perennial crops such as rubber, whose seeds were distributed by the colonial authorities from the early 20th century onwards, households modified the traditional land tenure system designed for hill paddy shifting cultivation. Most communities drifted away from a usufruct right conditional upon land plot utilization to form permanent individual property systems. Despite the lack of legal individual land titles, a household's land and perennial crops are secured through traditional institutions, local arrangements generally regarded as legitimate by the native population. Nevertheless, it is still debated how flexible and adaptable the traditional land tenure system is and, along with it, communal institutions in general (Cramb 1986, Cramb and Wills 1998, Doolittle 2004).

As a legacy of history and complex power relations, Iban native customary land developed into a micro-state shaped by socio-nature interactions and governed by distinct institutions (Ngidang 2006). Historically, even if state institutions loosely regulated the territorial extent of native communities, they had limited influence on native customary land, where traditional institutions governed most relations between environment and people. The situation, characterized by strong local institutions, led Sandin (1980) to describe native customary land institution as a fortress of Iban survival, ensuring economic stability as well as cultural security and continuity. Moreover, for Cramb and Wills (1990: 349) customary institutions are part of an "overarching community structure" as a prerequisite to sustain cooperative behaviours and social harmony (Cramb and Wills 1998: 57).

Challenging Native Customary Land Rights

Throughout its evolution, every Iban community had to negotiate the geographical extent of its customary territory with the colonial administration, and to justify it for subsistence needs. However, once secured within the borders of customary land, the community would retain control over its natural resource management institutions and property institutions. Recently, in the context of land hunger triggered by oil palm plantation expansion, this right has been challenged by state political power in new ways. During the colonial era, Iban people, as all indigenous communities, were granted native customary rights and territorial endowment along with a recognition of the primacy of their rights over Chinese migrants. Although the British legal system protected Iban traditional institutions pertaining to land management, it only recognized Western individual land titles as part of the legal property regime (and excluded the traditional Iban land system). According to the first agrarian legislations, the natives of Sarawak were licensees on Crown Land and by no means landowners (Ngidang 2005).

After Sarawak was formally ceded to the British Crown by the Brooke administration in 1946, the new government reasserted its power over land by implementing the Torrens land titling system in 1948. According to this system, developed in Australia, the survey of all individual plots was rendered mandatory as the government sought to create a uniform cadastral plan. However, customary land rights granted to Dayak and Iban communities, given the complexity of their tenure system, have been incompatible with the Torrens system. Until today, due to liberal and pluralist legal interpretation channels of the native land tenure system (Barney 2004: 332), individual land titling has been delayed in most Iban territories. However, through different rationales, the state has continually tried to circumscribe the Iban domain, intensifying regulations regarding access to land and its utilization. For example, the Forest Ordinance of 1953 restricted the expansion of hill paddy shifting cultivation beyond pre-existing locations. This curtailment was implemented alongside the designation of permanent forest areas: forestlands were in this way reserved for logging activities (Colchester 1993: 167).

The cornerstone of land regulations in Sarawak, the Land Code that came into effect in 1958, acknowledged pre-existing rights of Dayak communities but also imposed stricter conditions for legal recognition of customary land. The Land Code consolidated all previous land regulations and arbitrarily put an end to native territorial expansion. It stated that customary land rights could be granted only if the area claimed had been under continual use for given purposes prior to January 1958.[9] However, since Sarawak joined the Malaysian Federation in 1963, numerous amendments to the code have narrowed the definition of customary land, partially redefining native customary rights (Bulan 2006; Ngidang 2003, 2005; Hooker 1999).

In an attempt to re-regulate native land rights, the postcolonial government of Sarawak has enforced strict laws. In 2000 the clause according to which customary rights could be created "by any other lawful methods" was amended, consequently excluding recourse to cultural arguments to claim land rights. During the same year, an amendment ordinance enunciated that all disputed land would be presumed free of native customary rights until the contrary was proven.[10] Other amendments have increased the coercion power of the state. It is thus lawful, according to a 1999 clause, to extinguish native customary rights "upon direction issued by the Minister".[11] Although a provisional lease can only be issued on land deemed to be unencumbered by native customary rights, such leases have been issued to areas claimed by local communities but still without recognition by the state.[12] When provisional leases are issued on state land claimed by native communities, the law only requires publicizing the edict via a local newspaper six weeks before their coming into effect. Facing land dispossession, many communities have resorted to direct confrontation with oil palm company agents and paramilitary or police forces, and erected road blockades as the only way to hold back the companies and start legal proceedings. In this context, many cases of violence have been reported, showing the emergence of a resistance movement among indigenous people. Often, opponents of oil palm corporate expansion have been made politically aware and given networking means by local NGOs defending indigenous rights.

As all land is legally considered unencumbered by customary rights unless proof is provided by the community, the burden of proof—which can be costly—lies on indigenous claimants. If access to the judiciary is unrestricted, it is contingent upon available financial resources and effective mobilization as well as NGO support. In addition, groups of people challenging the state at the legal level can only hope for eventual political acknowledgment of their rights. Despite the landmark case of Nor anak Nyawai (2001), which helped to clarify the definition of native customary rights in terms of land access, many native communities are still entangled in a long-lasting struggle to obtain due recognition of their territorial claims (Bian 2001).[13] However, essential territorial elements of Iban and Dayak livelihoods are now legally recognized. Therefore, the community's territory (*pemakai menoa*), with its areas of cultivated land and fallow/secondary forest (*temuda*) and the community forest (*pulau galau*), have been included in the native customary rights' legal

Table 4.1 Sarawak. Extent of disputed native customary land (ha)

Division	Total NCR area in 1994 (area claimed + area recognized)	Area not recognized by the state and claimed as NCR land by native claimants in June 1994	NCR area recognized by the state in June 1994
Kuching	102,450	24,003	78,447
Samarahan	124,000	16,291	107,447
Sri Aman	174,192	35,723	138,469
Sibu	269,998	24,023	245,975
Sarikei	162,580	15,035	147,545
Kapit	475,255	11,098	446,157
Bintulu	82,000	5,224	76,776
Miri	201,063	19,703	181,360
Limbang	55,161	17,745	37,416
Total	1,628,699	168,845	1,459,854

Sources: Ngidang (2000: 34) citing Zainie 1994; Ministry of Land Development, Sarawak, 2006.

definition.[14] Native customary rights have thus been validated by the courts. But problems remain as the government of Sarawak is still issuing provisional leases on native customary land, such as in the upper Baram area.[15] The discrepancy between claimed territories by natives and recognized customary land by authorities is notable in most divisions of Sarawak. In fact, throughout the state, more than 160,000 hectares are still or were recently disputed. This figure represents the difference between customary land areas recognized by the state in the recent past (1994) and the area claimed and yet to be recognized legally as native customary rights land (NCR) (Table 4.1).

The 1988 and 1990 amendments to the Land Code were even more significant. According to these, private corporations, including foreign companies, are entitled to acquire native customary land for development purposes (Bulan 2006: 52; Ngidang 2005: 67). In order to facilitate large-scale expansion of commercial agriculture, the Land Custody and Development Authority was established in 1982. This institution is empowered to purchase or obtain provisional leases on state and native customary land for "private estate development", acting as an intermediary between landholders and private investors (Bulan 2006: 52). Moreover, further space was made for private investors now encouraged to develop native customary land. In accordance with decisions of the WTO Uruguay Round taken in 1993, changes brought to land regulations in Sarawak made way for international capital investments in an era of agrofood system restructuring (Thompson and Cowan 2000).

As it appears from these legislations, not only were conditions for obtaining recognition of native customary rights rendered stricter, but the power of the state to render unlawful native land utilization increased. The development strategy of the state based on oil palm plantation expansion was often interpreted by activists and many social scientists as a manoeuvre to take over native communities' resources. However, the state has constantly promoted large-scale oil palm expansion as a way to achieve rural development, in other words to enrich native communities and provide jobs. By the same token, some have argued that the purpose of these development measures was to strengthen patron-client relationships and reward political loyalty. While recognizing the new development policies as an attempt to improve the natives' living conditions, Ngidang (2005) emphasized that they were primarily aimed at enabling economic elites tied to political power to benefit from oil palm plantation activities. Amendments to the Land Code can be associated with a tighter regulation of land access and utilization in Sarawak. Land disputes are the key manifestation of a confrontational dynamic in which the state tries to reconfigure the rural economy through oil palm plantation expansion and customary land commoditization.

Case Studies

The case of two Iban communities facing land dispossession will now be examined in light of the processes that take place amidst the land hunger dynamic in Sarawak. Particular attention will be given to land claims put forward by the communities and the strategies used to cope with land dispossession threats. The study emphasizes the basis upon which rural dwellers legitimize their occupation of a territory and the claim laid upon it. It describes and analyzes how communities deploy strategies to counter the territorialization power of the state and its attempt to precipitate land commoditization for the expansion of oil palm plantation corporations. The cases offer insights into the way the relationship between the state's agencies and the communities evolved in the past decades. Powers and imperatives having determined the communities' territorial evolution are identified in order to discuss the consequences of new social interactions over natural resource management at the local scale.

The Geography of Livelihoods

The two communities studied here are located in the lowland area in Miri Division (Figure 4.1). The native community that for purposes of anonymity we shall call Sungai Sulasah is well connected with the main roads of the area. It is also near the Miri-Bintulu Highway, a segment of the main road in

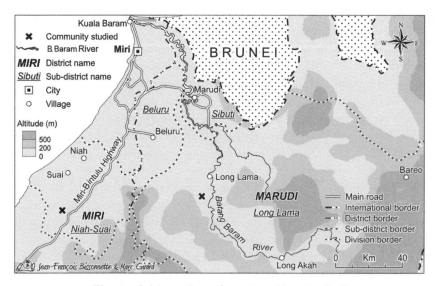

Figure 4.1 Location of communities studied

Sarawak. A number of corporate plantations are found nearby, mostly along the Miri-Bintulu road, with a palm oil extraction mill located in the vicinity of the community's territory. The community's proximity to the mill is such that villagers complain of water and air pollution problems from the mill.

Referred to as Sungai Binga, the other community is located farther east, more than 100 kilometres south of the estuarine village of Kuala Baram, on the Batang Baram River. More isolated than Sungai Sulasah, its only link to the paved highway is a logging road leading to a small service town. Otherwise, the boat service that serves populations scattered along the Batang Baram River represents the only means of transportation. From the town nearest to Sungai Binga, it takes approximately four hours to reach Marudi and six to reach Miri.

The geographical location of these communities impacts greatly on their economy, largely determining the commercial crops they can produce in relation to their market access. With different means, in markedly different contexts, most members of the two communities have been labouring to increase their income and improve their living conditions. Both these communities have long adopted rubber cultivation, but only members of the Sungai Sulasah community have been able in the recent past to start growing oil palm on their individual plots. Since this community is located near important marketplaces, smallholders of Sungai Sulasah can take advantage of market opportunities and easily sell in nearby markets crops such as pepper, oil palm bunches and fruits, as well as natural rubber sheets. Moreover, this location has granted several households better access to capital and, for men, opportunities to supplement the family income with off-farm activities such as driving and mechanical repair jobs in the automobile industry. During the period of the study, off-farm revenues from full-time or part-time activities were common among members of the Sungai Sulasah community, but much less so among members of the Sungai Binga community.[16]

Among the Sungai Binga community, rubber is now the only commercial crop cultivated, although many members also engage in off-farm jobs in the natural resource exploitation sector, such as fossil fuel extraction and logging. In the early 1990s, falling pepper prices coupled with sustained high input requirements forced all households to abandon cultivation of the crop. The facts that natural rubber can be stored and transported in sheets over greater distances and that cultivation requires limited care and input encouraged its adoption by most households. Paddy cultivation, along with gathering, fishing and hunting remain prominent features of the economy for both communities. However, the lesser extent of market opportunities in Sungai Binga causes a heavier reliance on shifting cultivation.

In both communities the state has intervened, mainly since the 1960s, to ensure the development of commercial agriculture activities on native customary land. On the Sungai Sulasah customary land, aside from the distribution of rubber seeds, government support through the Department of Agriculture mainly took the form of subsidies for the Creamy White Pepper Project.[17] This was a pepper production scheme implemented at the household plot level that accommodated pre-existing tenure. Subsidies for pepper cultivation from the Pepper Marketing Board flowed in until the 1990s, along with other sporadic financial aid for infrastructures such as basic running water facilities. During the 1980s, attempts on the part of the Department of Agriculture to initiate pepper cultivation in Sungai Binga were less successful and rapidly fell short, leaving rubber as the main source of income. Nevertheless, in Sungai Binga the diversity of incomes and food crop sources acts as a security against economic upheavals.

The Territoriality of Communities

The settlement of these two communities in the first half of the 20th century was sponsored and monitored by the state, which, from the beginning and acting as a trustee, controlled and regulated the territorial evolution in the pioneer phase of colonization. In Sungai Sulasah, settlement or colonization began after agreements were concluded with the adjacent Penan and Iban communities on the borders of the new customary land. People who were to form the Sungai Sulasah community were mainly from the Sungai Skrang area, although many families were recruited from the Bintulu and Lundu areas. There was geographical heterogeneity in the newly formed community, but everyone recognized the leadership of the headman from Sungai Skrang who had undertaken the settlement process. The Sungai Sulasah community in the 1950s, because of the greater population density in the chosen area, had to establish a clear boundary from the very first stages of its arrival. The Sungai Sulasah headman was in charge of discussing the community's boundaries with colonial officers.

In Sungai Sulasah, during the 1950s even perennial crop expansion on individual plots was closely supervised by government officials from the Department of Land and Survey.[18] In contrast to the Sungai Binga community, the Sungai Sulasah *pemakai menoa* frontier was surveyed and thus delineated with the help of government officials from Miri Division. By the end of 1954, the Niah-Suai sub-district officer had provided the community's headman with a map of customary land borders.[19] In the Sungai Binga case, due to reasons of low population density and relative remoteness, the territory claimed under

Table 4.2 Features of customary lands

Community	Population (no. of families)	Total area claimed (ha)	Area disputed (ha)	Customary land disputed (%)	Area allegedly infringed upon (ha)
Sungai Binga	~150	~12,000	~4,000	33	450
Sungai Sulasah	~80	3,200	~1,300	40	0

Source: Community mapping in Sungai Binga and Sungai Sulasah communities.

native customary rights was never fully surveyed by governmental authorities. This situation led to deleterious sporadic timber extraction during the 1960s by logging companies, but without any conspicuous resistance. The enforcement of regulations pertaining to land-use rights by colonial officers appears to have been much stricter in Sungai Sulasah than in Sungai Binga.

In accordance with Iban customs, both territories have been organized to allow each household to access plots located fairly close to the longhouses. In the colonization phase, usufruct rights over individual parcels were enforced in the two communities. Territories managed as corporate properties accessible only to members of the community (Appell 1995) have also been broadly delineated in order to conserve forest and water resources. Traditionally, institutions could guarantee sufficient land and natural resources to all households, although apparently not always on egalitarian terms.

In the Sungai Sulasah community, all accessible land resources were appropriated for shifting cultivation by the first generation of settlers, which formed a permanent use rights property regime.[20] All parcels first cleared were roughly delineated, and exclusive rights upon land were recognized to belong to households or individuals who had invested work on them. In order for every member of the community to access land, the head of a household had to divide his plots among the children willing to undertake agriculture, thus fragmenting land properties. If the fallow periods had to be reduced as a result of the redistribution process, the resource was still relatively abundant and the fallow period of shifting cultivation, ranging from 8 to 15 years, still apparently sustainable.

In the Sungai Binga customary land, geographical features created a slightly different organization pattern. Given the area available, here again the population on customary land divided among many longhouses enforced individual usufruct rights over most of the territory through the expansion of shifting cultivation. However, in this case a substantial area remained untouched by swidden activities and was further integrated into the community's livelihood as the *pulau*, the forest reserve. But unlike the previous case, Sungai Binga customary land still has an important land area dedicated to shifting cultivation under temporary usufruct rights or rotational cultivation rights, without well-defined plot boundaries.

The Iban communities under scrutiny here have been able to secure communal territories until recently. Their respective settlement or colonization process has involved state trusteeship. The paternalistic approach displayed by the developmental state has given rise to a system that respects native customs while extending state power. The state has engaged in redistributive strategies such as support for commercial agriculture diversification among the natives. Given their original conditions and land endowments, these communities have recently come to grips with the state and oil palm corporations in different ways. In fact, the recent state focus on corporate oil palm expansion in Sarawak has dramatically changed relationships between government agencies and Iban peasants.

Actions against Land Dispossession

In 1997, in both communities under study here, provisional leases were issued on lots representing substantial areas claimed as native customary rights land. For the Sungai Sulasah community, approximately 40 per cent of the native customary land was being theoretically ceded to the Sarawak Oil Palm company (SOP) by the Land and Survey Department (Table 4.2). Despite compensation offers formulated by the SOP, leaders of the community, seemingly with the full support of the population, decided to opt for juridical procedures in order to secure the whole area of their customary territory. In this case, households had dissimilar interests at stake in the dispute. In fact, the lot targeted by the private oil palm plantation overlapped only with some household plots and only one longhouse water catchment. Nonetheless, a consensus was seemingly reached among inhabitants regardless of personal stakes; they acknowledged the need to maintain the integrity of the territory of their community. Consequently, the headman of the community refused offers of compromise involving the establishment of a joint venture project on the community's customary land. Besides financial compensation, the headman was given an offer to join an oil palm plantation joint venture company on the native customary land of the community. However, this offer was associated with the dispossession threat and promptly refused by the headman and those who were consulted. In this case, the court case led by the villagers, along with the injunction enacted by the court of Miri, sufficed to push back the oil palm corporation. Calling upon the help of the NGO Borneo Research Institute of Malaysia and an independent lawyer, the villagers then launched legal proceedings.

The Sungai Binga community challenge was much more complex considering the various actors formulating overlapping claims on land and the dissent that existed within the community. In this case, the customary territory had been purposely delineated for integration into adjacent native customary lands, as part of a vast regional oil palm plantation scheme under the "Konsep Baru" development project. According to the Konsep Baru joint venture guidelines, native land rights would be transferred to a private corporation for a period of 60 to 90 years, during which native households would become shareholders in a process supervised by a state agency acting as a trustee of native landowners (Cf. Cramb, Chapter 3). Although the policy had been designed to allow negotiations among the community members as well as between the natives and the corporation to delineate areas of food crops, graveyards, etc., things turned out differently in practice. Power imbalances and the upper hand of the corporation led a substantial number of the inhabitants, who felt that they could not take part and control the transformation process, to categorically refuse the project. The Baram scheme, which served as a pilot development project in 1997 for the region, was perceived by many in Sungai Binga as less advantageous than previous socio-economic arrangements.[21] As the project was being put forward and the land surveyed, disputes surged within the community, causing inhabitants opposed to the scheme's implementation to break away from their longhouse to regroup into another one.

In Sungai Binga, many inhabitants were already weary of all redefinition of their land entitlements by the state. In this community, the claim for the recognition of native customary land had already been undermined by an encroachment of the oil palm company Loagan Bunut Plantation Berhad in the early 1990s. The company had then developed a plantation of more than 450 hectares on claimed customary land.[22] Swiftly mobilizing, a substantial number of community members engaged in actions to cope with the direct dispossession threat. In this struggle, legal NGOs brandishing rhetoric of indigenous inalienable rights supported the communities' claim.

In 1997, before legal procedures began, many members of the Sungai Binga community erected road blockades and drove out surveyors sent by the Land and Survey Department. As a result, dozens of people were arrested and allegedly brutalized. Besides engaging in direct confrontation, inhabitants addressed requests to the Miri High Court demanding immediate revocation of the logging licences. In the meantime, the main protagonists filed a court case to demand compensation for damages and obtain recognition of the full customary land claimed. The court's ordinance to the Forestry Department to halt logging operations eventually came into effect, but only after much land had already been deforested in preparation for the oil palm plantation. The involvement of several companies in the development scheme complicated the court case, but without preventing the construction of infrastructures such as an oil palm nursery within the native customary land.

Establishing the exact extent of native land being a prerequisite to setting up the scheme, land conversion operations have been delayed and will possibly be cancelled following the verdict pronounced in favour of the claimants by the court in October 2006.[23] In Sungai Binga some members of the community, allegedly influenced by the customary Iban leader of the sub-district, the *Penghulu*, signed individual contracts devolving their land rights to the newly formed corporate entity. The line between supporters and opponents of the project is, however, movable, which complicates the land claim process.

The two cases have important commonalities despite the different settings in which the land disputes took place. In both instances, political mobilization proved to be effective in protecting customary territory. However, the state was able to contest the claim of both Iban communities on densely forested territories, the *pulau*, by arguing that they were in fact idle, and therefore state land. Drawing on the jurisprudence case of Nor anak Nyawai (2001), the communities have been striving to demonstrate before the court the indispensable role of their reserve forest. In fact, for the Sungai Sulasah community, the forest reserve serves as a water catchment as well as a source for collection of forest products, mineral resources exploitation, and recreation. It is managed by the community through its communal institutions as a private corporate property.[24]

Legitimizing Native Land Rights

In Sungai Binga, members of the community have resorted to modifying their land management by extending the cultivated area to secure a territory contested by a plantation company. Many inhabitants, using their knowledge of land laws, which forbid disturbing native territories planted with crops, have readily put under cultivation areas directly threatened by logging operations. This strategy of physical occupation of claimed land has helped the population to demonstrate actual use of coveted land for basic food needs. For the Sungai Binga community, it has succeeded in countering further land appropriation by the exogenous oil palm company.

Numerous protected species of trees were planted by the villagers to physically delineate customary land. By virtue of laws regulating forest exploitation, it is formally forbidden to cut down certain tree species traditionally used by Dayak populations. These species (rattan, illipe nut, etc.) were promptly planted by Iban villagers in areas disputed by the logging company clearing the way for the oil palm scheme. Considering the requirements of the juridical apparatus, landscape modification practices like this one have been observed among many native communities of the Global South (Unruh 2006) and constitute a means to legitimize land use. These practices would not have been possible without the legal knowledge disseminated by local NGOs through workshops and information sessions.

Although, unlike Sungai Binga, the Sungai Sulasah community was never directly physically threatened by logging companies and their bulldozers, it adopted a similar strategy to legitimize its land claim. In Sungai Sulasah, smallholding oil palm cultivation serves as a means to assert landownership. It also represents an endogenous economic development strategy. Members of the Sungai Sulasah community are aware, as are those of the Sungai Binga, that putting areas under cultivation constitutes a means to legitimize their control. In doing so, members of the Sungai Sulasah not only increase their revenue, but they also adopt an economic development strategy that is considered legitimate and modern in official discourses of the state. These discourses emphasize the key role of commercial perennial crop expansion, such as oil palm, for economic development. Therefore, many other communities rely on the strategy of adopting smallholding oil palm cultivation.[25] In this context, many Iban smallholders are able to argue that corporate oil palm plantations on native customary land through joint ventures are not needed, since smallholders themselves can deploy their own resources and achieve the same result.

Although the greater reliance on oil palm cultivation by Iban smallholders in Sungai Sulasah is not a direct outcome of the struggle against land dispossession, many can make sense of this strategy as an assertion of economic independence. The inhabitants of Sungai Sulasah maintained that they were entitled to remain in control of the disputed area since they sought to use it for what was allegedly the same purpose as that of the corporate plantation: growing oil palm to achieve economic development. A significant number of heads of household interviewed (8 out of 25) were planning to or were already relying exclusively on oil palm to meet their economic needs.[26] This tendency shows a move towards increased dependence on oil palm cultivation, which constitutes a break away from traditional ways of dealing with risk as well as from previous peasant diversified economic strategies encompassing subsistence agriculture. As a result, securing land rights becomes imperative for new Iban farmer-entrepreneurs who seek to expand their personal oil palm plantation.

These cases exemplify opposition to plantation corporation expansion on native customary land despite offers of monetary compensation or shares in joint ventures. The factors explaining the resolute opposition of communities to relinquishing territorial assets to oil palm plantation corporations can be found in the attachment to pre-existing livelihood strategies. Moreover, many native peasants and scholars consider that plantation expansion eradicates the direct link between peasant households and territorial-economic management (Songan 1993, Ngidang 2006, Unruh 2006). This direct link would be replaced by an abstract one, as economic means of production would be mediated by an exogenous private company (Cramb and Wills 1990). It comes as no surprise that some communities and individuals choose to remain in control of their local economy and able to seize changing market opportunities. Relying on family labour and individual managerial arrangements is often preferred. In addition, the customary model confers greater possibilities for food security and the preservation of a landscape imbued with cultural significance for indigenous communities.

In this context of confrontation, members of the two communities interviewed expressed a fierce conviction in their land rights while recognizing their vulnerability. Recent events have often reinforced a feeling of marginalization among Iban peasant populations, nurturing a pride in economic autonomy and self-help (Ngidang 2006). In both cases, despite limited access to capital and lack of formal property titles, the communities opted for an endogenous initiative to achieve economic development based on agriculture, without even requesting legal individual land titles.

The Significance of Community Mapping

The involvement of NGOs in native customary land claims was briefly mentioned above. However, community mapping, a specific feature of the involvement of NGOs specializing in native land rights, deserves further attention. Community mapping is the main tool used in Sarawak to support land claims against oil palm plantation corporations and the state's Land and Survey Department. Mapping tools, traditionally a monopoly of the state and its technocratic apparatus, play an indispensable role in enforcing territorial regulations (Vandergeest and Peluso 1995). The relative democratization of cartographic technologies has allowed some NGOs to empower communities in challenging the territorialization power of the state (Peluso 1995). Therefore, community mapping acts as a form of counter-mapping instrument, a countervailing strategy to present before juridical instances an alternative representation of a customary territory. Politically aware members of native groups who revert to community mapping defy the state's hegemonic means to represent and regulate territories (Majid-Cooke 2003).

The NGOs Borneo Research Institute of Malaysia and Sahabat Alam Malaysia were involved respectively with the villagers of Sungai Sulasah and Sungai Binga. These organizations mainly provided legal training for the community leaders with whom they were dealing. But more important, the NGOs provided community mapping services. In both communities, community mapping was used to delineate the boundaries of the territory claimed as customary land, the *pemakai menoa*. Not only were the borders of a community's customary land determined with a GPS, but also household plots and collective properties were surveyed using the same method. However, in the two communities, the survey of the longest individual plots and of the one involving the most delicate negotiations between members of the communities lagged behind the survey of the forested area (*pulau*). Community mapping may have important repercussions on the outcome of court cases about customary land rights only if its legitimacy can be recognized by the government of Sarawak, even though its validity has already been alleged by the court.[27]

Besides the legal importance of community mapping, the delineation of individual land plots has implications for the evolution of the Iban land tenure in Sarawak. Although Iban customary laws have always recognized individual rights over land, borders often remained flexible and usufruct rights were traditionally devolved when a plot was abandoned by an individual or a household. Community mapping may further contribute to the disappearance of the flexible tenure, which started with the extensive adoption of perennial crops. Cramb and Wills (1990: 350) highlight that traditional Iban institutions allow land resource redistribution in case of a highly uneven access structure. However, the pressure of oil palm plantation on land resources might irremediably erase this possibility. In fact, the threat of land dispossession from oil palm corporations has forced the community of the Sungai Binga, for instance, to ascribe a greater legitimacy to land utilization by clearly identifying the function of each plot. In this case, community mapping as a response to the pressure exerted on land may contribute to the disappearance of mechanisms of land redistribution. In Sungai Binga abundant land resources allow for maintaining a form of common land bank. However, in this case the boundaries of valuable plots are being clearly demarcated through community mapping and clearly distinguished from the area where land could be redistributed.

Conclusion

This study has sought to shed light on some of the manifestations of power struggles surrounding contemporary native customary rights in Sarawak. It provides examples of mobilization of Iban communities against oil palm corporate infringement upon territories governed by customary institutions. And it situates this confrontational dynamic in the broader context of oil palm corporate expansion, legal restrictions on native customary rights, and state development projects.

Communities under scrutiny are defending customary territories against real and potential encroachment from oil palm corporations. The communities reject corporate oil palm plantation expansion on their customary territories in order to pursue their own agrarian livelihood strategies. Influential members of these communities perceive corporate oil palm encroachment as a threat to the continuity of customary natural resource management systems. They consider that this encroachment on their land has the potential to compromise the economic development of the community. Moreover, many native communities hesitate to relinquish customary property institutions; this explains in part the opposition to the joint venture development project of the Konsep Baru.

For people struggling against encroachment and unwanted interventions from exogenous oil palm corporations on their customary land, land rights are considered as the basis of economic control and flexible agricultural production. As Ngidang (2006) argues, land in Sarawak is perceived by Iban rural communities as a security, an asset on which to fall back in case of economic hardship. Furthermore, the threat of land dispossession on a community can trigger effective mobilization of its members if the local political dynamic allows it. High levels of social cohesiveness already existing in this agrarian setting can be strengthened when livelihood strategies are perceived to be endangered.

Customary land claims are likely to remain key issues in rural Sarawak. The widespread rejection of the Konsep Baru policy in the countryside of Sarawak could eventually lead to the creation of rural development orientations that would be more widely accepted in the state. However, land claims brought before the court will remain a dominant issue, delaying state agenda and nurturing confrontation, physically or rhetorically. However, as customary land rights are ultimately being defined by the state government and its power to legislate, customary land claims will remain a political issue and not strictly a juridical one. As long as the state administration fails to accommodate reasonable native customary land claims, Iban and native peasant actions and forms of mobilization will continue to play an important role in agrarian change in Sarawak.

Notes

1. This research was carried out as part of a master's programme in geography, completed at the Université de Montréal in 2007. It was funded by the Fonds québécois de recherche sur la société et la culture (FQRSC) and the "Challenges of the Agrarian Transition in Southeast Asia" (ChATSEA) research project. It greatly

benefited from the project's conceptual approach and overall logistics, summarized in De Koninck (2004), as well as from advice from several of its members.

2. The three forestry plantation projects approved in the sole Bakun Catchment Area cover a total of 320,000 hectares. For environmental sustainability reasons, Sahabat Alam Malaysia is very critical of the projects. (Sahabat Alam Malaysia, Sarawak, Rengah Sarawak online, 21 June 2007).

3. In 2006, the 271,000 hectares gazetted for oil palm plantations came under the Ministry of Land Development targets (500,000 hectares for 2010) (Kuching, Ministry of Land Development, 6 Sept. 2006).

4. Although important tracts of land have been alienated for joint oil palm plantation ventures, the survey of customary land and individual plots, along with the high level of coordination required between different state institutions—namely, the Sarawak Planning Unit, the Ministry of Resource Planning, the Ministry of Land Development, the Land and Survey Department—often greatly delays the land titling process.

5. For more information on the policy and its implications, see Majid-Cooke 2002, 2003; Ngidang 2002.

6. Ministry of Land Development, Sarawak, government's view expressed by YB Tan Sri Datuk Alfred Jabu anak Numpang, "Handbook on New Concept of Development on Native Customary Rights Land", p. 10.

7. Personal interview with Bian Baru, Kuching, 5 Sept. 2006. Other sources refer to as many as 200 communities (The Borneo Project, *Legal Aid Fund for 2009*, http://borneoproject.org/article.php?id=686, accessed 28 Jan. 2009).

8. Personal interview with Jok Jau, Sahabat Alam Malaysia, Marudi Division, Sarawak, 6 July 2007.

9. Section 5(2) of the Sarawak Land Code, 1999: Native customary rights can be established by: (a) felling of virgin jungle and occupation of the land; (b) planting of land fruit trees; (c) occupation or cultivation of land; (d) use of land for burial ground or shrine; (e) use of land of any class for rights of way; and (f) other lawful methods.

10. Ordinance 2000, Section 7(A)(3)(b).

11. Sarawak Land Code, 1999, Part II(3).

12. Thanks to Dr Robert Cramb for clarifying this particular legal mechanism of provisional lease issuance.

13. Landmark case of Nor anak Nyawai & 3 ors. v. Borneo Pulp Plantations Sdn Bhd & 2 Ors.

14. Rayuan sivil No Q-01-42-2001, Judge, Court of Appeal, Dato' Haji Hashim bin Dato' Haji Yusoff.

15. Sahabat Alam Malaysia is involved in other native customary land claims in the upper Baram region, such as in the Ubra communities. Personal interview with anonymous informant related to Sahabat Alam Malaysia, Marudi Division, Sarawak, July 2007.

16. Qualitative data were gathered through brief surveys among focus groups in the communities during July and August 2006. This statement excludes people considered as community members who live outside the community.

17. On this occasion, Rumah L received RM250,000 and Rumah T RM200,000. In the same way, other subsidies were allocated for pepper cultivation by the government agency the Pepper Marketing Board (PMB) (Legal statement, suit No. 2x-xx-1998).

18. Legal statement: In the High Court in Sabah and Sarawak at Miri, suit No. 2x-xx-1998 (MR) Rumah L and Rumah T claimants v. Superintendent of Lands & Surveys Miri Division, Government of Sarawak and Sarawak Oil Palm Berhad.

19. Ibid.

20. In the High Court in Sabah and Sarawak at Miri, suit No. 2x-xx-1998 (MR) Rumah L and Rumah T claimants v. Superintendent of Lands & Surveys Miri Division, Government of Sarawak and Sarawak Oil Palm Berhad.

21. The Baram oil palm scheme was initiated in February 1997. It included 550 households in 14 longhouses forming a joint venture with the company Perlis Plantation Berhad and the Sarawak Land Development Board (SLDB). The company withdrew from the project in early 2002 (Bulan 2006: 54).

22. Sahabat Alam Malaysia, Sungai Binga community mapping results.

23. Personal interview, Sahabat Alam Malaysia, Marudi Division, Sarawak, 6 July 2007.

24. Appell (1995) establishes the difference between open access and multiple ownerships. They should not be confused, the latter being regulated by people regrouped to manage the property in a corporate system.

25. Data based on a survey carried out in the sub-districts of Niah, Tinjar and Bakong with 253 respondents by Ngidang and Majid-Cooke (2006). Results show that 90 per cent of the cultivators surveyed were involved in oil palm cultivation, this activity representing their first occupation.

26. In the community there were about 80 households. Among the sample of 25 heads of household, which totalled 129 individuals, 20 were involved in oil palm cultivation on their own parcel, although 8 were already relying or were about to rely solely on oil palm cultivation. In many cases, household incomes were supplemented by the employment of older children in nearby private oil palm plantations.

27. Sahabat Alam Malaysia, Marudi, personal interview, 13 Aug. 2006; "With the passing of the Land Surveyors Ordinance 2002, the combined effect of Sections 20 and 23 entail that a person who is not a licensed surveyor from the Land and Survey Department cannot make, authorise or sign any cadastral map" (Bulan 2006: 61).

References

Aoki, M. and Y. Hayami, eds., *Communities and Markets in Economic Development*. Oxford: Oxford University Press, 2001.

Appell, G.N., "Community Resources in Borneo: Failure of the Concept of Common Property and Its Implications for the Conservation of Forest Resources and the Protection of Indigenous Land Rights", in *Local Heritage in the Changing Tropics, Innovative Strategies for Natural Resource Management and Control*, ed. Greg Dicum. *Yale School of Forestry and Environmental Studies Bulletin* 98 (1995): 2–56.

Barney, K., "Re-encountering Resistance: Plantation Activism and Smallholder Production in Thailand and Sarawak, Malaysia", *Asia Pacific Viewpoint* 45, no. 3 (2004): 325–39.

Bian, B., Summary of the landmark case of Nor anak Nyawai & 3 ors. v. Borneo Pulp Plantations Sdn Bhd & 2 Ors, 2001.

Brookfield, H. and Y. Byron, "Deforestation and Timber Extraction in Borneo and the Malay Peninsula: The Record since 1965", *Global Environmental Change* 1, no. 1 (1990): 42–56.

Bulan, R., "Native Customary Land: The Trust as a Device for Land Development", in *State, Communities and Forests in Contemporary Borneo*, Asia-Pacific Environment Monograph 1, ed. F. Majid-Cooke. Canberra: ANU E Press, 2006, pp. 45–64.

Case, W., "Malaysia: New Reforms, Old Continuities, Tense Ambiguities", *The Journal of Development Studies* 41, no. 2 (2005): 284–309.

Colchester, M., "Pirates, Squatters and Poachers: The Political Ecology of Dispossession of the Native Peoples of Sarawak", *Global Ecology and Biogeography Letters* 3 (1993): 158–79.

Cramb, R.A., "The Evolution of Iban Land Tenure", Working Paper No. 39. Melbourne: Centre of Southeast Asian Studies, Monash University, 1986.

———, "Problems of State-Sponsored Land Schemes for Small Farmers: The Case of Sarawak, Malaysia", *Pacific Viewpoint* 33, no. 1 (1992): 58–78.

Cramb, R.A. and I.R. Wills, "The Role of Traditional Institutions in Rural Development: Community-based Land Tenure and Government Land Policy in Sarawak, Malaysia", *World Development* 18, no. 3 (1990): 347–60.

———, "Private Property, Common Property, and Collective Choice: The Evolution of Iban Land Tenure Institutions", *Borneo Research Bulletin* 29 (1998): 57–70.

De Koninck, R., "Challenges of the Agrarian Transition in Southeast Asia", *Labour, Capital and Society* 37 (2004): 285–8.

Doolittle, A., "Colliding Discourses: Western Land Laws and Native Customary Rights in North Borneo, 1881–1918", *Journal of Southeast Asian Studies* 34, no. 1 (2003): 97–126.

———, "Powerful Persuasions: The Language of Property and Politics in Sabah 1881–1996", *Modern Asian Studies* 38, no. 4 (2004): 821–58.

Drummond, I. and D. Taylor, "Forest Utilisation in Sarawak, Malaysia: A Case of Sustaining the Unsustainable", *Singapore Journal of Tropical Geography* 18, no. 2 (1997): 141–62.

Harvey, D., *A Brief History of Neoliberalism*. Oxford and New York: Oxford University Press, 2005.

Hayami, Y., "Communities and Markets for Rural Development under Globalization: A Perspective from Villages in Asia", Foundation for Advanced Studies on International Development, GRIPS / FASID Joint Graduate Program, 2006.

Hooker, M.B., "A Note on Native Land Tenure in Sarawak", *Borneo Research Bulletin* 30 (1999): 28–40.

Kerkvliet, B.J., *The Power of Everyday Politics: How Vietnamese Peasants Transformed National Policy*. Ithaca and London: Cornell University Press, 2005.

King, V.T., "'*Politik Pembangunan*': The Political Economy of Rainforest Exploitation and Development in Sarawak, East Malaysia", *Global Ecology and Biogeography Letters* 3 (1993): 235–44.

Li, T.M., "Articulating Indigenous Identity in Indonesia: Resource Politics and the Tribal Slot", *Society for Comparative Study of Society and History* 42, no. 1 (2000): 149–79.

———, "Engaging Simplifications: Community Based Resource Management, Market Processes and State Agendas in Upland Southeast Asia", *World Development* 30, no. 2 (2002a): 265–83.

———, "Local Histories, Global Markets: Cocoa and Class in Upland Sulawesi", *Development and Change* 33, no. 3 (2002b): 415–37.

———, *The Will to Improve: Governmentality, Development and the Practice of Politics.* Durham and London: Duke University Press, 2007.

Majid-Cooke, F., "The Politics of 'Sustainability' in Sarawak", *Journal of Contemporary Asia* 27, no. 2 (1997): 217–41.

———, "Vulnerability, Control and Oil Palm in Sarawak: Globalization and a New Era?" *Development and Change* 33, no. 2 (2002): 189–211.

———, "Maps and Counter Maps: Globalised Imaginings and Local Realities of Sarawak's Plantation Agriculture", *Journal of Southeast Asian Studies* 34, no. 2 (2003): 265–84.

Malaysia government, *Yearbook of Statistics.* Kuching, Sarawak: Department of Statistics, 2006.

Martin, P.M., "Comparative Topographies of Neoliberalism in Mexico", *Environment and Planning A* 37, 2 (2005): 203–20.

Ngidang, D., "Native Customary Land Rights, Public Policy, Land Reform and Plantation Development in Sarawak", *Borneo Review* 8, no. 1 (1997): 63–80.

———, "People, Land and Development: Iban Culture at the Crossroads", in *Iban Culture and Development in the New Reality*, ed. D. Ngidang, S.E. Sanggin and R.M. Saleh. Iban Cultural Seminar, 1998; Dayak Cultural Foundation, 2000.

———, "Contradiction in Land Development Schemes: The Case of Joint Ventures in Sarawak, Malaysia", *Asia Pacific Viewpoint* 43, 2 (2002): 157–80.

———, "Transformation of the Iban Land Use System in Post Independence Sarawak", *Borneo Research Bulletin* 34 (2003): 62–78.

———, "Deconstruction and Reconstruction of Native Customary Land Tenure in Sarawak", *Southeast Asian Studies* 43, no. 1 (2005): 47–75.

———, "Cultural Landscape, Market and Land Development Policy in Sarawak", unpublished final version, University of Malaysia, Sarawak, 2006.

Ngidang, D. and F. Majid-Cooke, "Self-empowerment among Iban Oil Palm Smallholders in Miri Division, Sarawak", draft, unpublished, 2006.

Osman, S., "Globalization and Democratization: The Response of the Indigenous Peoples of Sarawak", *Third World Quarterly* 21, no. 6 (2000): 977–88.

Peluso N.L., "Whose Woods Are These? Counter-mapping Forest Territories in Kalimantan, Indonesia", *Antipode* 27, no. 1 (1995): 383–406.

Pringle, R., *Rajahs and Rebels: The Ibans of Sarawak under Brooke Rule, 1841–1941.* Ithaca: Cornell University Press, 1970.

Puyok, A., "Voting Pattern and Issues in the 2006 Sarawak State Assembly Election in the Ba' Kelalan Constituency", *Asian Journal of Political Science* 14, no. 2 (2006): 212–28.

Reinhardt, J., "Administrative Policy and Practice in Sarawak: Continuity and Change under the Brookes", *Journal of Asian Studies* 29, no. 4 (1970): 851–62.

Sandin, B., *Iban and Adat Augury*. Penang, Malaysia, for School of Comparative Social Sciences, 1980.

Sarawak, Laws of, *Land Code Chapter 81*. Kuching: State Attorney-General's Chambers, 1999.

Sarawak, Ministry of Land Development, *Handbook on New Concept of Development on Native Customary Rights (NCR) Land*. Petra Jaya, Kuching, 1997.

Sarawak, Ministry of Land Development and Ministry of Rural Development, *Briefing to Pasukan Projek Penggubelan Dasar Tanah Negara Jabatan Ketua Pengarah Tanah & Galian by Hj. Chaiti Bolhassan*, 10 Feb. 2006.

Songan, P., "A Naturalistic Inquiry into Participation of the Iban Peasants in the Land Development Project in the Kalaka and Saribas Districts, Sarawak, Malaysia", *Borneo Research Bulletin* 25 (1993): 101–22.

Sutlive Jr., V.H., *The Iban of Sarawak, Chronicle of a Vanishing World*. Virginia, IL: Waveland Press, 1988.

Thompson, S.J. and J.T. Cowan, "Globalizing Agro-food Systems in Asia: An Introduction", *World Development* 28, no. 3 (2000): 401–8.

Unruh, J.D., "Land Tenure and the 'Evidence Landscape' in Developing Countries", *Annals of the Association of Geographers* 96, no. 4 (2006): 754–72.

Vandergeest, P. and N.L. Peluso, "Territorialization and State Power in Thailand", *Theory and Society* 24, no. 3 (1995): 385–426.

5
Oil Palm Plantations in Sabah: Agricultural Expansion for Whom?

Stéphane Bernard and Jean-François Bissonnette
University of Ottawa and University of Toronto

Understanding the Territorial Expansion of Oil Palm in Sabah

In the state of Sabah, oil palm became the most important cash crop in terms of area and production (tons) in the early 1980s. This rapid growth was attributable both to the numerous commercial outlets for the crop and to its high profitability in an increasingly favourable market. Moreover, the Asian financial crisis of 1997–98 was beneficial to the development of this production. Along with the entire Malaysian agricultural sector, the oil palm economy showed resilience, pursuing its growth in the midst of the crisis while several other sectors suffered a significant decline (Cheng Hai 2000). World consumption trends, such as increasing demand for so-called green energies, namely, agrofuels, also contributed to oil palm growth in Southeast Asia and particularly in Sabah (Carter *et al.* 2007, Wicke 2008), as Malaysia undertook to become a world leader in oil palm-based agrofuel technologies and production.[1]

In Sabah, the conversion of vast tracts of forested and fallow land to agricultural use has therefore occurred in response to the increasing global demand for both agrofuel and edible oil. Consequently, the state, which until recently remained at the periphery of the capitalist economy—notwithstanding the extensive logging operations to which it was subjected after World War II—is currently being reconfigured through agricultural expansion and as a result becoming further integrated into global and national productive spaces. Taking into account the major socio-environmental changes brought about by forest exploitation, we wish to emphasize the new implications and consequences of the oil palm crop boom. This is particularly important because, unlike in the case of log extraction, the clearing of large forest areas for export-oriented intensive production of a monoculture leads to a new regime of landownership and access, as well as entirely new labour relations.

In this regard, Sabah can be seen as a site of intense transformation where major agrarian and environmental changes are induced by contemporary oil palm expansion (see Chapter 1). This chapter constitutes a contribution to the study of transformations induced by contemporary oil palm expansion in the state. We argue that the colonial capitalist plantation model still informs current oil palm plantation expansion and that it primarily serves the interests of capitalist expansion represented by private corporations. These transformations involve new actors that nevertheless operate, as we argue, in continuity with colonial actors and the broader historical context of estate expansion in Sabah. In fact, the gradual withdrawal of the state as the main agent of agricultural expansion at the turn of the 1990s and its replacement by large corporations has rendered the dynamic of oil palm expansion more problematic in Sabah. This process has caused profound transformations not only for Sabahan smallholders' livelihoods but also for an increasingly important population of migrants from neighbouring countries who have settled in rural areas. Just like in neighbouring Sarawak (Cf. Chapter 4), a large number of land disputes have also taken place in Sabah because of the expansion of commercial oil palm estates. Consequently, as estate expansion accelerates, competition for land between private concerns and communities also increases.

The aim of this chapter is therefore to look into the socio-economic and territorial transformations induced by oil palm expansion in Sabah.[2] Who benefits from oil palm expansion in Sabah? What are the dynamics that drive this large-scale territorial transformation process? To tackle these questions, we adopt a comprehensive approach, integrating the historical emergence of plantation agriculture in colonial Sabah with the study of contemporary actors, whether state, national or global, in oil palm plantation expansion, while focusing on some of the consequences of oil palm plantation expansion on the environment and on smallholders.

In the first section of the chapter, we look into the classical plantation model. Based on empirical material, the second and third sections underline the continuity between the colonial and contemporary periods while the fourth section addresses the recent shift from social objectives to global market imperatives in postcolonial Malaysian agricultural expansion policies. In the fifth and sixth sections we examine major contemporary actors in oil palm plantation expansion—global capital and foreign labour—and in the seventh and eighth sections we explore the contradictions inherent in the roles of the state, which are to maximize development, provide a legislative basis for environmental management and facilitate economic growth through plantation agriculture. This is followed by a case study of the Kinabatangan River Valley, one of the areas most affected by oil palm expansion. This case study illustrates the complexity and highly problematic nature of oil palm expansion for the Orang Sungai as well as Indonesian migrants (mainly Bugis), often considered "nonofficial residents". These populations often find themselves caught between oil palm estate expansion and the extension of forest and wildlife conservation areas.

Back to the Classical Plantation Model

By 1999 the state of Sabah was already the main producer of crude palm oil (CPO) in Malaysia, with 25.3 per cent of the total, just ahead of Johor. During recent years, the dominance of Sabah as the prime producer of palm oil among Malaysian states has been increasing. By 2007, with close to 5.6 million tons of CPO, Sabah's share of national production had reached 35 per cent (MPOB 2008), on a total cultivated area of approximately 1.3 million hectares. This represents 30 per cent of Malaysia's planted area and almost 90 per cent of all cultivated land in Sabah (or 17 per cent of total land). Despite the presence of oil palm smallholders and the rapid emergence of fruit and vegetable market gardening for local urban and export markets, Sabah's agricultural landscape is clearly dominated by oil palm estates. In general these estates, or rather plantations, each cover more than 700 hectares and rely on a specific management model.

According to several authors (Courtenay 1965, Beckford 2000, Goldthorpe 1987), a plantation is characterized by a highly specialized form of production, usually single-crop farming, and is export-oriented. It is therefore by essence integrated into the global trade system. Moreover, such a large-scale exploitation model relies on tight labour control. The labour force, a fundamental feature of the plantation, is embedded in the highly centralized management operations of the plantation (Beckford 2000). Sabah's oil palm plantations thus belong to an agricultural system present throughout the humid tropics. However, such a system is not defined by the nature of its production—oil palm is also grown by smallholders—but rather by the scale of its operations, the bureaucratic structure of management and specific forms of labour control. As for standard smallholding agriculture, it relies on family labour and management; it also mobilizes local financial resources and adapts to local needs and diversified economic practices (Goldthorpe 1987, Hayami 2002).

Plantations and smallholdings represent two highly contrasted agricultural management models. The former are essentially oriented towards profit maximization, while the latter respond, beyond the need for profitability, to imperatives of household socio-economic development. For the purpose of this study, we associate independent smallholdings with agricultural development schemes, such as those initiated by the Federal Land Development Authority of Malaysia (FELDA) in 1956. FELDA was established by the federal government in order to confer a social role to large-scale agricultural schemes. Landless peasants or farmers with unsustainable land-size plots were sponsored to join large agricultural schemes. These were meant to favour the growth of a prosperous smallholding sector within a wider structure managed by the agency (Bahrin and Lee 1988). However, in Sabah, areas converted to agriculture strictly for the socio-economic development of the local peasantry remained quite limited in size. Rather, the capitalist plantation model introduced under the rule of the British North Borneo Chartered Company (BNBCC) was never really put aside and is thriving once again.

Along with worldwide liberalization of investment and trade, demand for oil palm in world markets has spurred a new plantation boom. Derived from the colonial era, this particular mode of agricultural expansion is obviously leaving an important footprint on Sabah's landscape. In addition, the contribution of the plantation sector to local socio-economic development appears questionable. In fact, most of the labour as well as management and capital involved in the plantation economy originate from outside of Sabah, while a substantial proportion of the benefits leave the territory. Nevertheless, it would be wrong to look at the plantation as an entity completely isolated from the rest of the territory in which it evolves. As pointed out by Hayami (2002), plantations are often considered the best way to extract monetary value from lands left idle. One question remains: Who benefits from the conversion of idle land into oil palm plantations—as if land was ever idle—and at what cost to local populations and the environment?

The Colonial Background of Contemporary Agricultural Expansion

Oil palm plantation expansion in Sabah shows elements of continuity with the colonial era. In fact, the current agricultural expansion phase was made possible by a Western land regulation system. Established in 1881 by the BNBCC, it was meant to facilitate the commoditization of land and resources. In order to extract maximum revenue from its investments in Sabah, the BNBCC was empowered by the British government to set up a system that would facilitate natural resource exploitation. Plantation expansion on so-called unused territory was supervised by an enterprise and its management apparatus, which in turn relied on the provision of capital and labour (Beckford 2000, Hayami 2002). From the late 19th century onwards, numerous estates were thus allocated to European interests by the BNBCC. This process constituted an efficient colonization tool that transformed much of Sabah into a new frontier for commercial agricultural development.

By virtue of the charter granted by the Crown, the BNBCC had the abiding responsibility to respect the rights and customs of the various local population groups, such as the Kadazan-Dusun and the Murut. As a consequence, these populations were kept isolated from all exogenous development initiatives, which instead relied mainly on foreign labour. In fact, colonial British North Borneo called upon Chinese and Javanese to provide labour on the plantations (Cleary 1992; Doolittle 2003, 2004).

The establishment of a Western legal system based on individual property and the administration of land access and taxation was a prerequisite for the provision of new agricultural territories to European investors (Cleary 1992). In parallel, indigenous territories were given legal recognition as the administration enforced native land rights. This policy symbolized the paternalistic attitude of the company's administration towards North Borneo's ethnic groups and illustrated clearly its attempt to establish a dual economy. Not only would the colonial economy based on mining and plantation activities be restricted mainly to non-natives, but the Kadazan-Dusun as well as the other groups' domain would be protected from external influences. However, in reality the temporary and non-adjacent nature of the Sabahan groups' agricultural plots, on which they practised shifting cultivation, made it difficult for them to obtain due recognition of their territorial rights (Doolittle 2004). The implementation of a Western-type land tenure system quickly turned out to be incompatible with the customary systems of the Kadazan-Dusun and Murut. Although most traditional land tenure systems were reshaped by colonial intervention, an important proportion of the population was nevertheless able to obtain ownership rights on individual parcels or customary lands (Appell 1985, Doolittle 2004).

By the end of the 19th century, advantageous investment conditions and low levels of land taxation had attracted a large number of investors (Doolittle 2003). From 1930 to 1976, the state of Sabah referred to the Land Ordinance of 1930 as the unique land management guideline. This Land Code was established in order to regulate private property on alienated land. It considered all land free of native customary land status, thus rendering it no longer accessible to indigenous people. The land property framework did not involve any planning strategy to open lands to cash crops, although by 1970 the cultivated area had already reached 313,000 hectares (Jomo *et al.* 2004). Lawless agricultural expansion thus occurred, especially on the east coast of the state, where cocoa plantations were to flourish eventually. Despite the fact that legal arrangements were inadequate, they allowed for the delineation and gazetting of native customary land, the backlash of which was that the remaining portion of the land was granted for future plantation expansion.[3]

The resulting dual economy contributed to the economic marginalization of a significant proportion of local population groups (Cleary 1992, Doolittle 2003). Moreover, the Land Code of 1883 and subsequent land ordinances such as the Ladang Ordinance of 1913 resulted in a landownership system with the explicit function of circumscribing the extent of shifting cultivation, deemed as environmentally destructive.

Over recent decades, in the neighbouring state of Sarawak the ambiguity of native land rights has given rise to numerous conflicts surrounding land access and utilization, resulting from oil palm corporations' land grabbing, itself facilitated by state regulations (Majid-Cooke 2002, Ngidang 2005). In Sabah, if direct confrontations between communities and private corporations seeking to grab native land are not as common as in Sarawak, they nevertheless do occur throughout the countryside.[4] In the central and isolated district of Tongod, Murut indigenous communities have been victims of forced resettlement and land dispossession. Along with unsound practices, the implementation of the Forest Management Policy privatizing logging extraction activities and forest management has caused forced resettlement in interior areas of Sabah. Without having to extinguish native customary rights, companies were still able to grab people's land. Conflicts in many districts often stemmed from contradictory interpretations of the Land Ordinance of 1996 (Long *et al.* 2003). Following the landmark case of the Rumah Nor in Sarawak, several communities in Sabah have also been trying to obtain land titles.[5] However, the outcomes of legal struggles remain uncertain. Thus, in continuity with colonial administration, postcolonial development policies are still aimed at expanding cash crops and suppressing shifting cultivation wherever it can be found (Lim and Douglas 1998, Doolittle 2004).[6] According to interviews with local development practitioners, a significant proportion of Sabahan ethnic groups have abandoned shifting cultivation, although it is still practised in numerous areas of the east coast, such as in the vicinity of the Crocker Range National Park, and in central areas.[7] However, most swidden cultivation is now done within delineated native territories legally recognized by the state or upon which claims were laid in compliance with the 1953 Land Laws (Doolittle 2004). The application procedure to obtain land titles is lengthy, due to the insufficient institutional capacity of the state. In 2007 there were more than

100,000 land alienation applications still pending, due also to overlapping claims between households.[8] Nevertheless, despite the seemingly peaceful land claim process in Sabah, the land system configuration—with its native reserves enforced during colonial times and other systems to settle Sabahan populations and confine their land utilization—has allowed private interests to gain land-use rights to vast areas. In this manner, extensive unclaimed areas are legally delivered to the private sector for oil palm plantation expansion.

From Social Objectives of Agricultural Expansion to Global Market Imperatives

Under the BNBCC rule, cash crop plantations were the main instrument of territorial transformation in Sabah. Then, from 1946 to 1963, the British colonial administration that took over after the surrender of the Japanese army developed an in situ agricultural programme to support native populations. Renamed Sabah, British North Borneo joined the Malaysian Federation in 1963, and cash-crop expansion was intensified. In fact, the new constitutional status of Sabah allowed federal agencies such as FELDA and the Federal Land Consolidation and Rehabilitation Authority (FELCRA) to extend their operations into Sabah territory. While most agencies started their operations

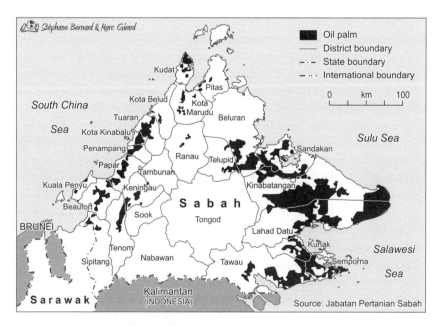

Figure 5.1 Sabah. Oil palm-planted area, 1997

with cocoa and rubber tree crops, with an important proportion under a smallholding regime, oil palm estates rapidly became predominant in the 1980s, replacing most other perennial crops as the latter's prices were tumbling down in export markets. That was the case on the east coast of Sabah, particularly in the Kinabatangan region, where cocoa was gradually replaced by oil palm estates. This has had important consequences on land tenure for the smallholders involved.

Although the private sector held 74 per cent of oil palm plantations in 2007 (Table 5.1), FELDA played a decisive role in inducing a high concentration of oil palm cultivation on the east coast of Sabah (Figure 5.1, Table 5.2). The impressive palm oil production capacity of that region is mainly due to FELDA, which was granted 143,000 hectares in 1979 (Bahrin and Lee 1988). Most of this land was turned over to agricultural use between 1985 and 1995 (Sutton 2001). Smallholders recruited by federal agencies were allocated a cash crop plot already integrated into a large-scale production structure. Linked by production contracts to the public-owned company, these smallholders benefited from the agency's processing structures and market access. The agency also provided every smallholder with chemical inputs, pest-control technologies and so forth.

Table 5.1 Sabah. Oil palm-planted area by type of ownership, 1980–2007 (ha)

Type of property	Private estates	FELDA	FELCRA	State schemes (SLDB)	Small-holders (licensed)	Total
1980	60,840	0	0	27,900	1,260	90,000
%	68	0	0	31	1	100
1990	132,906	88,506	0	48,477	6,282	276,171
%	48	32	0	18	2	100
1995	323,487	116,083	0	60,529	18,034	518,133
%	62	22	0	12	4	100
1999	704,393	124,642	4,682	67,345	40,260	941,322
%	75	13	1	7	4	100
2003	831,330	124,799	13,840	98,555	66,777	1,135,301
%	73	11	1	9	6	100
2007	949,407	113,874	14,690	94,087	106,186	1,278,244
%	74	9	1	7	8	100

Sources: Cheng Hai 2000, Palm Oil Registration and Licensing Authority (PORLA), Malaysian Palm Oil Board (MPOB) 2008.

Table 5.2 Sabah. Oil palm-planted area by district, 2002–5

District	Area of oil palm, 2002 (ha)	% of total district area, 2002	Area of oil palm, 2005 (ha)	% of total district area, 2005	Growth 2002–5 (ha)	Growth 2002–5 (%)	Total land area by district (ha)
Tawau	96,977	15.8	119,082	19.4	22,105	18.6	6,125,000
Semporna	43,115	37.7	48,109	42.0	4,994	10.3	1,145,000
Lahad Datu	205,810	27.7	234,757	31.5	28,947	12.3	7,444,000
Kunak	65,387	57.7	71,398	63.0	6,011	8.4	1,134,000
Sandakan	99,944	44.1	106,635	47.1	6,691	6.3	2,266,000
Kinabatangan	299,385	45.3	303,943	46.0	4,558	15.0	6,605,000
Tongod	3,818	0.4	20,638	2.1	16,820	81.0	10,054,000
Beluran[a]	168,589	21.8	259,052	33.6	66,904	33.0	7,719,000
Kudat[b]	3,293	2.6	5,824	4.5	2,382	73.6	1,287,000
Pitas	3,803	2.7	3,721	2.6	-82	-2.2	1,419,000
Kota Marudu	5,174	2.7	5,214	2.7	40	0.8	1,917,000
Kota Belud	394	0.3	415	0.3	21	5.0	1,386,000
Ranau	350	0.1	400	0.1	50	12.5	3,608,000
Tuaran	413	0.4	413	0.4	0	0.0	1,166,000
Kota Kinabalu	…	…	…	…	…	…	351,000
Penampang	…	…	5	…	5	100.0	466,000

Table 5.2 (continued)

District	Area of oil palm, 2002 (ha)	% of total district area, 2002	Area of oil palm, 2005 (ha)	% of total district area, 2005	Growth 2002–5 (ha)	Growth 2002–5 (%)	Total land area by district (ha)
Papar	3,077	2.5	3,463	2.8	386	11.1	1,243,000
Beaufort	13,740	7.9	19,673	11.3	5,933	30.2	1,735,000
Sipitang	367	0.1	432	0.2	65	15.0	2,732,000
Kuala Penyu	1,113	2.5	1,147	2.5	34	3.0	453,000
Tenom	4,273	1.8	4,815	2.0	542	11.3	2,409,000
Keningau[c]	9,599	2.7	17,808	5.0	-1	0.0	3,533,000
Tambunan	25	…	45	…	20	44.4	1,347,000
Nabawan[d]	33	…	1,415	0.2	1,382	97.7	6,089,000
Total	1,028,679	…	1,228,404	…	199,725	16.3	73,633,000

Notes: a: Total oil palm area and land district area include Telupid District as shown in Figures 5.1 and 5.2; b: Total Kudat oil palm area and land district area include Matunggong District (not shown in Figures 5.1 and 5.2); c: Total oil palm area and land district area include Sook District as shown in Figures 5.1 and 5.2; d: Formerly named Pensiangan District.

Source: Jabatan Pertanian Sabah. Soil and Survey Department and Sabah Yearbook of Statistics, 2006.

The objective of the scheme was to grant landownership to settlers once they had completely reimbursed the loan obtained to start up production. The overall objective was to create a new class of prosperous landowners for whom agricultural production externalities would first be internalized by the state. Especially in Peninsular Malaysia, during the 1980s and 1990s, FELDA became a dominant agent of agricultural expansion.[9] On the Sabah side, the Sabah Land Development Board (SLDB) was established in 1976. Replicating FELDA's development model, it concentrated its efforts on opening new agricultural schemes for the settlement of impoverished peasants (Jomo et al. 2004). Furthermore, independent smallholders without state support were able to secure a modest but significant share of the total oil palm crop area—8 per cent of Sabah's total planted area in 2007 (Table 5.1). However, smallholders must be considered in many respects as an epiphenomenon of plantation and state scheme development, for they are dependent upon palm oil extraction and transportation infrastructures provided by powerful companies. Only plantation companies and state agencies are able to internalize the costs of major infrastructure development (Hayami 2002). Transportation infrastructure as well as processing and shipping facilities provided by either private or state concerns are indispensable to the large-scale expansion of oil palm cultivation.

By 1990, FELDA and SLDB were responsible for managing an oil palm area as extensive as the one managed by the entire private estate sector. However, since then the situation has rapidly evolved in favour of the latter (Table 5.1), the implementation of the Sixth Malaysia Plan (1991–95) representing a turning point in state agricultural development. Gradually, both federal and state governments ceased to subsidize agricultural expansion. In addition, in Sabah most FELDA operations have been privatized. Consequently, existing schemes no longer pursue the goal of improving the lot of impoverished rural dwellers. Finally, between 1993 and 1997 a number of FELDA schemes were regrouped, their total number falling from 92 to 75 (Sutton 2001). Through centralization of operations, the prime objective is now to maximize profits, which involves relying on foreign workers to overcome the shortage in local labour.

Since the very beginning of its operations in Sabah, FELDA has encountered difficulties in recruiting local settlers for its schemes and, to meet agricultural expansion targets, has had to call upon migrant workers with legal as well as illegal status (Fold 2000, Sutton 2001, Jomo et al. 2004). Consequently, the 54 schemes established during the 1990s have since been transformed to a point where they are basically managed in the manner of traditional plantations. Contrary to the situation that prevailed in the

Malaysian Peninsula during the first decades following independence, most local communities in Sabah had access to land, although it was not recognized through individual ownership titles. As the initial FELDA schemes were laid out, opportunities to join the urban economy were already undermining smallholder agriculture expansion and rural resettlement. Thus, by the end of the 1990s, on the east coast of Sabah almost 95 per cent of workers on FELDA's plantations were migrants from the Philippines or Indonesia, despite the agency's initial objective of providing rural Sabahans with opportunities for socio-economic improvement (Sutton 2001). The fact that the agency has ceased to open new schemes speaks for itself, indicating its decision to consolidate its operations and to leave agricultural expansion in the hands of the private sector, in line with the recommendations formulated in the Sixth Malaysia Plan.

The expansion of oil palm cultivation has indeed largely been taken over by the private sector. By the mid-1970s, important financial incentives were already offered to attract private investments in agriculture. A variety of means to reduce taxes and allow entrepreneurs to reinvest in agricultural development were proposed. By the early 1990s, negotiations within the General Agreement on Tariffs and Trade following the Uruguay Round and the creation of the World Trade Organization (WTO) led to further liberalization of trade in agricultural products (Thompson and Cowan 2000). This new international development strategy and regulation regime had repercussions on national policies, reshaping the rural realm while favouring the control of the agro-industrial sector by global capitalism. The Third National Agricultural Policy (1998–2010) confirmed and reinforced this orientation by encouraging competitiveness and productivity within the private sector. It did not, however, prevent smallholders from responding to the global market demand and substantially increasing their share of oil palm cultivation. On the contrary, Sabah state agricultural development policies support intensification and commoditization of production among the smallholder population (Government of Sabah 1999).

The ineffective attempt at resettling peasants on crop schemes for socio-economic improvement abruptly ended in the early 2000s. Palm oil production in Sabah is now possible only if it can compete, in terms of production costs and quality, with other oil crops in the global markets. This also means expanding the planted area and alienating large tracts of land to a much greater extent than needed for the improvement of the livelihoods of the local inhabitants. Social considerations have therefore been largely pushed aside, especially since the early 1990s, as oil palm plantations increasingly embody trade liberalization in the food economy (Sutton 2001, Thompson and Cowan

2000). However, some small-scale relocation schemes have been implemented to solve specific livelihood problems for populations facing issues caused by forest conservation initiatives as well as expansion of oil palm estates.

Actors in Global and National Capitalism at the Local Scale

As in colonial times, the international agribusiness sector imports technologies, capital and labour in a territory designed as an "enclave" of capitalist production. Contemporary planters in Sabah are corporate entities based mainly in the Peninsula and representing Malaysian patrimonial capitalism. By patrimonial capitalism, we refer to foreign companies that were overtaken by national interests when Malaysia gained independence. Many of these companies, such as Sime Darby, Golden Hope and Kumpulan Guthrie, are now major actors in oil palm plantation expansion in Malaysia and Sabah.[10] These entities succeeded in carving out, at low cost, their exclusionary enclaves of agricultural production in Sabah and throughout the Southeast Asian archipelago. More than 40 per cent of the shares of these corporations are controlled by the Malaysian National Equity Corporation, locally known under the name of Permodalan Nasional Berhad (PNB) (*The Star*, 2006). The PNB itself constitutes a division of the Bumiputera Investment Foundation, a financial institution established by the state in 1970 to pursue the New Economic Policy and thus increase the capacity of the federal state to produce new sources of wealth for the Bumiputera population.[11] Along with these companies partially controlled by the state, other companies, essentially private, originate from Peninsular Malaysia. The latter corporations owe their investment capabilities in Sabah to the historical development of the highly lucrative plantation economy in Peninsular Malaysia, the Malaya of British colonial days.

After the foundation of the Federation of Malaysia in 1963, the plantation tycoons of former Malaya gradually expanded their operations into Sabah and then beyond as plantations were deployed all over the Malay Archipelago, notably in Sumatra, Kalimantan and Mindanao (Bernard 2006). In other words, Southeast Asia's oil palm boom was seemingly led by Malaysian tycoons. In fact, many important corporations operating in Sabah—such as PPB Oil (with 51,135 hectares in Sabah in January 2007),[12] Consolidated Berhad, Asiatic Development and IOI corporation—maintain their headquarters in Peninsular Malaysia and are listed on the Bursa Malaysia (formerly the Kuala Lumpur Stock Exchange), with several of these "Malaysian" corporations pursuing their expansion into Indonesia and beyond, for example, in the Solomon Islands, Africa (Nigeria) and South America (Colombia, Venezuela and Brazil). In addition, major oil palm plantation companies from Sabah, such as IJM Plantations and Hap Seng, have expanded their operations to Indonesia. Both these companies are listed on the Bursa Malaysia. Among the prominent Malaysian corporations active in Sabah, KL Kepong, Highlands and Lowlands are also listed on the London Stock Exchange, while United Plantations is listed on the Copenhagen Stock Exchange.

Within global capitalist expansion, the oil palm plantation sector represents an interesting capital outlet to provide investors and shareholders with sustained growth. In this regard, local Sabahan government executives alleged that approximately 80 per cent of oil palm plantations in Sabah were owned by foreign interests (Bernama 2003). In the economic landscape created by the latest phase of globalization, not only have oil palm companies expanded outside their traditional areas, but they have diversified their activities. Among the corporate entities mentioned, many are now involved in real estate and construction as well as in various activities linked to the transformation of palm oil. The globalized productive order in which these palm oil production enclaves function, as Ferguson (2006) puts it, is characterized by transnational networks connecting plantations to their capital sources "in a point to point fashion" (Ferguson 2006: 40). In sum, ventures in oil palm plantation expansion, once listed on the stock exchange, are open to global capital flows and are regulated according to the imperatives of profit maximization, competition and accumulation (Wood 2002).

Foreign Workforce in Plantations

As was the case during the colonial era, labour engaged in agriculture in Sabah comes overwhelmingly from abroad. In fact, migrants from neighbouring countries, essentially from the Philippines and Indonesia, make up the majority of workers on plantations (Hugo 1993, Sutton 2001, Liow 2003, Sadiq 2005). According to the 2000 Malaysian census, non-citizen residents represent about a quarter of the 2.679 million inhabitants of the state. This figure does not take into account the allegedly substantial "illegal" immigrant population. This being said, some state officials have adopted a tolerant, almost *laissez-faire* attitude towards unregulated immigration, this continuous flow of migrant labour having become essential to the local economy. While the many checkpoints set up on roads are supposedly aimed at intercepting illegal immigrants, they do not affect the situation much. In reality, the presence of illegal migrant workers is tolerated if not somehow tacitly encouraged. This has been the source of a constant dispute between the federal government and Sabah's authorities regarding illegal migrants in Sabah. In fact, Malaysian laws that forbade the use of migrant workers during the economic decline caused by the Asian financial crisis of 1997–98 did not apply to the agricultural and domestic work realms (Liow 2003). Despite the August 2002 ultimatum that migrant workers without a permit would have to leave the state or else incur the risk of being imprisoned and deported, their number still amounted to 500,000 in July 2006, according to the state's estimate (*The Borneo Post*, 16 July 2006). Although this number gives an idea of the extent of the phenomenon, many allege that the state is purposely underestimating it. Moreover, the situation of illegal migrant workers generally remains precarious, especially during economic downturns. In 2002, shanty towns built on the outskirts of Sandakan by illegal immigrants were destroyed.[13]

Labour input from surrounding countries where the cost of living is lower remains vital to the plantation sector in Sabah. Entrepreneurs recruiting labour often succeed in employing workers without a permit. Indonesian migrant workers are usually prized as they speak a language well understood in Malaysia and accept low wages (Liow 2003). Studies tend to demonstrate the existence of established migration networks from the Indonesian islands to Sabah. Most immigrants seem to come from the eastern part of the Indonesian archipelago, along an axis stretching from Flores to Sulawesi and Kalimantan (Hugo 1993). Manifestly, this migrant labour force, which comes from a population basin characterized by poverty, exerts a downward pressure on wages in the plantation economy of Sabah. This situation undermines the attraction potential of the sector for local rural inhabitants, despite the enduring high poverty rate registered among them. Average daily wages for unqualified day labourers on plantations—usually ranging from RM10 to RM15 (although wages as low as RM7 per day were reported in Kinabatangan District until 2005)—fall below the Malaysian poverty threshold. This is thus a sector that exploits disparities between national economies while perpetuating poverty and remaining unappealing to local peasants who still have access to land.

The State Government: Victim or Instigator of Agricultural Expansion?

Since the British North Borneo Chartered Company was established, political and economic factors have together reinforced the function of Sabah as a territory specialized in natural resource exploitation. In 2000, industrial production represented 12.7 per cent of GDP, against 33.4 per cent for the whole of Malaysia (Government of Malaysia 2003). In addition, from 1991 to 2000 Sabah's GDP growth rate was the lowest in the federation (Government of Malaysia 2003). By 2000, despite a substantial reduction, the rate of poverty was still the highest in the federation. According to official estimates from the federal government, in 2005, 16 per cent of the population was considered to be living in poverty, with the rate as high as 35 per cent in rural areas.[14] According to economists, this indicator points to the structural nature of problems inherent to the Sabahan economy (Teck Wai 2001). Poverty problems go back to the colonial period, when Sabah was relegated to the periphery. Since then the state has maintained a relationship of dependence with the industrial core of the federation, still centred on the west coast of Peninsular Malaysia.

More recent economic development policies have been aimed at fostering diversification and an increase in the share of manufactured products in the GDP. Federal development policies have also fostered rural infrastructure development programmes. This development orientation lies at the core of the Eighth Malaysia Plan (2001–5) and Third National Agricultural Policy (1998–2010) (NAP3). However, as in the case of Peninsular Malaysia, the portion of state aid allocated to industrial development depends largely on state revenues derived from the primary sector. In fact, revenues from natural resource exploitation contributed nearly 80 per cent of the state revenue in 2006 (Sabah State Budget Speech 2007). In addition, deficiencies in land tax collection deprive the state of substantial revenues. Finally, Sabah's slow industrial take-off obviously limits the contribution of the secondary and tertiary sectors to the state budget. This state of affairs, along with the economic growth objectives of the state, as enunciated in federal policies, partly explains the renewed pressure on agricultural expansion.[15]

The relative importance of agricultural activities increases along with the substantial shrinking of the forestry sector's contribution to state revenue, itself partially attributable to overexploitation of the resource. This contribution, which dropped from approximately 70 per cent during the 1970s to less than 20 per cent in 2001 (Juin *et al.* 2000), continues to lose ground. In contrast, state revenues derived from sales tax on palm oil are booming, along with their relative importance in Sabah's economy. According to 2006 estimates, the contribution of taxes on crude palm oil to the total state revenue was estimated at almost 27 per cent (RM157 million) (Sabah State Budget Speech 2007). It is the most important contribution from a single product to the revenues of the government. To this must be added 4 per cent of state revenue coming from land taxes and land lease premiums for oil palm cultivation.

In Malaysia, palm oil production is predominantly export-oriented, with about 90 per cent of palm oil exported in the 2000s (MPOB 2008). In 2004, palm oil exports constituted 33.7 per cent of all export revenues in Sabah—ahead of petroleum, which accounted for 22.3 per cent (IDS 2005). Despite the importance of revenues originating from oil palm, the latter's overall economic role is less obvious. Until the end of the 1990s, the existing 58 palm oil mills in Sabah were extracting only crude oils. These had to be exported to Peninsular Malaysia and other international markets to undergo additional transformations (IDS 2005).

However, the palm oil transformation industry has evolved rapidly since the establishment of the Lahad Datu Palm Oil Industrial Cluster (POIC) in 2006. The initial phase of the POIC in Lahad Datu consists, as of 2010, of a scheme of 460 hectares equipped with infrastructure that provides a base for high-technology industries for the transformation of palm oil. The site also harbours industries related to the production of fertilizers. According to the development plans laid down originally, an important biodiesel production centre using palm oil as raw material should be in operation by now. In view of the success of the initial phase, which would have attracted RM1.8 billion in domestic and foreign investments, the POIC authorities have undertaken the second phase of the project and propose to more than double the area of the cluster in Lahad Datu in the coming years (Bangkuai 2010).

Many stakeholders still predict that the trickle down from these major biofuel projects into the local economy will remain modest.[16] Beyond the heavy reliance on inputs, equipment and labour, the expatriation of profits is likely to be such that benefits for the local population will remain meagre. Meaningful spillover from converting vast areas to monocropping is still to be seen, despite the substantial contribution of sales tax on palm oil to the state budget. Overall, 17 per cent of the Sabahan territory is allocated solely to this production, which exerts major pressures on the environment, the overall impact of which is still to be fully assessed. Table 5.2 shows the particularly dramatic pace and intensity of oil palm expansion that occurred in most of Sabah's districts from 2002 to 2005; by 2009 there was still no evidence that this trend had slowed down. According to findings of field trips in the districts of Nabawan, Sook, Keningau and Tambunan in June 2009, oil palm expansion activities are continuing and involve road upgrading, logging, land clearing, and new oil palm nurseries and plantings. This seems to be accompanied by a gradual privatization of land assets, with smallholders remaining marginal in terms of the territory they occupy.

Developmentalism, Monocropping and Environment

The conversion of hundreds of thousands of hectares of forested areas for oil palm plantation expansion is the result of development policies legitimized by the imperative of economic growth and modernization. Otherwise, it seems that the amount of land needed to address poverty problems for the Sabahan population would have been much less. In this regard, regulations established at the federal level since the 1990s emphasized the need to maximize natural resource utilization. Most soils endowed with some agricultural potential and even less fertile soils were allocated to agriculture. In addition, the National Forestry Act of 1984 created Permanent Forest Estates (PFE) in order to favour logging operations and agricultural expansion. Areas not allocated to conservation purposes (12.3 per cent of the territory) nor to commercial forest for exploitation under the PFE (36.4 per cent of the territory) were commoditized and thus integrated into the state's land market (IDS 2005). All alienated lands located outside urban areas, recognized native lands or part of the PFE, are likely to be converted into plantations—whether for wood production (timber plantations) or for oil palm (Figure 5.2). Land is usually granted under renewable provisional leases that bind the legal holder to cultivate the area within a period of five years in order to avoid land speculation. The apparent rationality of environmental management, sustained by the development rhetoric of state and federal governments, allows oil palm plantation expansion to continue in Sabah.

Furthermore, agricultural expansion in Sabah is taking place in a context where environmental impact assessment studies are often ignored or partially overlooked (Sutton 2001). The Land Capability Classification (LCC) remains the main planning tool to achieve land-use management. As such, the LCC

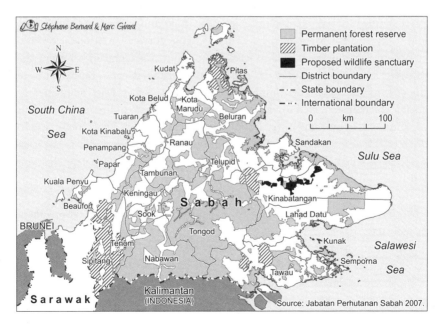

Figure 5.2 Sabah.The permanent forest estate, 2007

was implemented in 1976 with the purpose of regulating the allocation of new agricultural land. In this way, the LCC allowed for the identification of territories suitable for agriculture (Jomo *et al.* 2004). However, by 1997 almost 20 per cent of all cultivated land was located outside areas suitable for agriculture according to the LCC. If the extent of the impact of these transformations has yet to be fully scrutinized, it is likely that inherent limits to environmental sustainability will be eventually trespassed and the stability of biophysical environments compromised (Fitzherbert *et al.* 2008).

Although logging operations did turn out to be deleterious for the environment in the past (Jomo *et al.* 2004, Brookfield and Byron 1990), expanding cultivated areas, mostly for plantation monocropping, has even more significant repercussions. In fact, plantation development entails the extension of a land management model that transforms an original ecosystem for a long-term period. Deforestation and agricultural expansion are designated as the first cause of biological impoverishment in the area. The two activities are complementary in many aspects: plantations are generally laid down on land previously deforested and criss-crossed by logging roads (Kummer and Turner 1994, McMorrow and Talip 2001). What appears alarming is the conversion of dipterocarp forests—which are endowed with one of the highest floristic concentrations on Earth—into monocropping. These forests shelter 10 per cent of all known floral species, including some 265 to 390 species of dipterocarp, a tree family especially prized in forestry for its commercial value (McMorrow and Talip 2001). Moreover, lands left barren between deforestation and agricultural activities are subject to erosion and nutrient run-off from the topsoil (Hartemink 2005). In addition, the rugged topography adds to the risk of excessive environmental degradation. Substantial quantities of chemical fertilizers, used to obtain high yields, flow into bodies of water. This constitutes a significant source of pollution that adversely affects populations, often to the extent of poisoning the traditional food supply chain, as has been reported in Kinabatangan District (see case study below). Ultimately, chemical fertilizer run-off into seawater is likely to severely harm marine ecosystems in the surrounding seas (De Vantier *et al.* 2004). It is in the perspective of environmental sustainability that the expansion of oil palm plantations must be assessed.

The east coast of Sabah, still mostly forested until the 1970s, became the state's key area for public and private plantation development (Figure 5.1, Table 5.2). This was due to several factors, among them the presence of generally good soil and extremely low population densities. This situation was said to be partially attributable to the presence of the Crocker Range, which impeded people's movement from the more populated western part of the state to the extended narrow alluvial plain on the east coast. Until the establishment of the British North Borneo Chartered Company in Sabah, only a few human settlements based on sea trade were dispersed along the shoreline (Morgan 2001). This situation led the authorities to implement land regulations that would facilitate large-scale agricultural expansion in this "almost empty" area. According to the Palm Oil Registration and Licensing Authority (PORLA 2005), Tongod and Kinabatangan Districts in Sabah still had some 500,000 hectares available for commercial plantation agriculture (Figure 5.1, Figure 5.2, Table 5.2). This has led to a drastic transformation of the landscape, particularly noticeable in the Kinabatangan region.

The Case of the Kinabatangan River Valley[17]

Some of the largest oil palm estates in Sabah can be found in what is most currently referred to as the Lower Kinabatangan River Basin. At the centre of the region we find: (1) intensive logging operations; (2) expanding oil palm estates; (3) increasing conservationist pressure and expansion of preservation areas. These processes have important consequences for the local population, including the Orang Sungai and Indonesian immigrants of Bugis background, who already have to adapt to challenging livelihood conditions and are squeezed or encircled by either oil palm estates or conservation areas. In fact, it is estimated by local authorities and researchers that migrants mainly from Indonesia and the Philippines make up approximately three-quarters of the population in Kinabatangan District. Most of these migrants are currently working on oil palm estates.

A land-use map (Figure 5.3) of the Kinabatangan River Basin illustrates the impacts of recent oil palm estate expansion on local communities and their territories and contextualizes the argument presented previously in this chapter. The map combines information from satellite images, topographical maps and field notes. Like other areas of Sabah's east coast and interior, Kinabatangan District provides an example of an agricultural frontier that expanded rapidly over the past decades, gradually transforming territories that used to have a high biological diversity.

Kinabatangan District consists of a flat plain never surpassing 20 metres above sea level. In the 1970s it was dominated by cocoa cultivation, initially developed by smallholders. But by the 1980s this area had become the centre of oil palm estate expansion in Sabah, setting the pace for the transformation of ecological and human landscapes.

Until the mid-1970s, the Lower Kinabatangan River Basin was almost entirely forested and harboured a highly diversified flora and fauna. The region

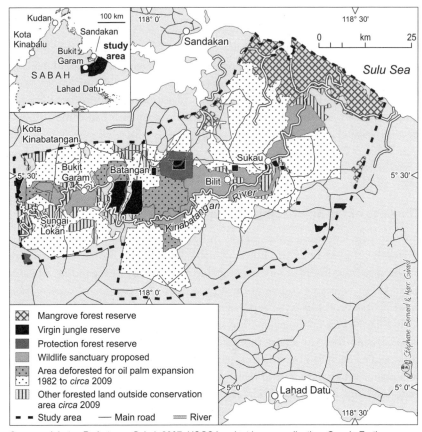

Sources: Jabatan Perhutanan Sabah 2007, USGS Landsat images collection, Google Earth;
        The Digital Chart of the World

Figure 5.3 Kinabatangan. Oil palm expansion, deforestation and
conservation, 1982-2009 (*circa*)

is inhabited by, among several groups, the Orang Sungai, known in Sabah for occupying and living along the riverbanks, hence their name, which literally means the "river people".

The analysis of a series of topographical maps, going back to 30–50 years ago and covering the entire Kinabatangan Basin, reveals traces of previous land and resource utilization regimes. These topographical maps reveal a mosaic of diversified forest formations, logging mills, camps and roads, small plantations, shifting cultivation patches of different sizes, and small settlements. These have gradually vanished under the new regime of oil palm plantation expansion. The land requisitioned for this type of agricultural expansion was too often considered "empty" by land developers, or "actors in global and national capitalism". In fact, over the past 30 years logging and the subsequent clearing of land for the expansion of monocultures have had dramatic impacts on biodiversity. But more important, oil palm plantation expansion has led to far-reaching transformations in land access and labour relations. It has also contributed to the transformation of the livelihoods of the Orang Sungai, a group that used to live under a different regime of resource utilization.

In order to evaluate the retreat of the forest in the context of oil palm expansion in the Kinabatangan, a study area of approximately 4,000 square kilometres was set in the Lower Kinabatangan, where large fields of oil palm trees have been planted. It is also the region where some of the new conservation areas have been established or planned over the past decades. Satellite images for 1982 and 2009 were analyzed, and they reveal an impressive loss of forest cover: exactly 1,657 square kilometres, most of which have been converted to oil palm (Figure 5.3). Even half of the riverbank forest, a particularly fragile component of the ecosystem, has made way for oil palm crops.

The planning of a wildlife sanctuary around the Kinabatangan River, advocated by environmental NGOs, was a response to this process of land-use conversion by oil palm corporations. But several oil palm plantation lobbies succeeded in postponing beyond 2008 the final gazetting of an ecologically diverse area along the banks of the Kinabatangan River (Figure 5.3). By virtue of the Sabah Wildlife Conservation Enactment of 1997, a substantial portion of the water catchments of this river was to be protected. But despite continual downscaling of conservation, the full implementation of the wildlife sanctuary has yet to occur. Nevertheless, despite the work of locally based NGOs financed by international donor agencies, environmental conservation in the context of the Kinabatangan Wildlife Sanctuary also involves continual partnership with local inhabitants. The challenge therefore lies in convincing the residents of villages adjacent to the projected wildlife sanctuary not to clear their own plots of forest. In fact, for villagers it seems that the potentially high returns to be obtained from oil palm cultivation offset the benefits they can expect from environmental conservation. As a result, a large number of villagers have converted their plots to oil palm, a reaction that can be viewed as a form of resistance to official environmental conservationism. Currently, the landscape in the Kinabatangan consists of a thin stretch of preserved forest along the Kinabatangan River, as most forest conservation areas are now almost completely surrounded by oil palm trees (Figure 5.3).

In addition, the negative effect of official environmental conservation is that rural dwellers usually devolve their conservation responsibility to specialized agencies. According to our evaluation, in 2009 approximately 30 per cent of the entire surface dedicated to conservation in the Kinabatangan River Valley had been transformed into oil palm fields, the most affected area being the wildlife sanctuary area, where 34 per cent of the forest reserve had been transformed over the period 1982–2008. Six per cent of the Protection Forest Reserve had suffered from conversion to oil palm. When interviewed in 2007, officials of the Sabah Wildlife and Forestry Department admitted to being overwhelmed at times by the scale of the task of protecting Sabah's remaining forest. According to members of Partners of Community Organisations in Sabah and the Kinabatangan Orang Utan Conservation Project, both of which have been involved in environmental conservation for more than 20 years in the Kinabatangan area, governmental participation in environmental conservation remains quite limited.

According to interviews carried out by researchers in various locations of Kinabatangan District, oil palm plantation expansion exerts a significant pressure on local communities' land resources. Peoples such as the Orang Sungai and other non-citizen residents such as Indonesians (Bugis) and Filipinos employed in the oil palm economy have complained of their lack of land. Many of those communities can be seen as literally encircled by oil palm plantations. In fact, the entire area has been affected by a double regime of deforestation and agricultural expansion by estates as well as by smallholders. The Orang Sungai living there have been particularly affected by conservation initiatives as well as by oil palm expansion. This new regime of land utilization has also brought new immigrants to the area. To benefit from the infrastructure laid down by commercial estates, the Orang Sungai have also developed their own oil palm plots.

Field notes collected by Lesley Potter in 2005 concerning the small town of Bukit Garam in the middle of Kinabatangan District (Figure 5.3) provide details on the process at the village level and complement what has been observed through the interpretation of topo maps and satellite images. In 2005 the main activity in the area was oil palm cultivation on estates, with Indonesian labourers and their families living in barracks. Interspersed between the estates were houses of smallholders, predominantly Orang Sungai. During the 1970s these had been encouraged by the government to move from farther upriver to this logged-over area, each family being allocated 15 acres of land. At that time there were no roads or schools, and the people upriver were quite isolated, earning a living by collecting forest products (including bird's nests), fishing and swidden farming. Prior to planting oil palm, most families had grown cocoa, with planting material initially provided by the Department of Agriculture.[18] Some families with large houses along the main road had acquired more than 30 acres and were reasonably well off. They mentioned that the area could not accommodate any new settlers.

One farm that was surveyed in detail consisted of 35 acres, with only 11 actually planted with oil palm. The farmer bought the forested land in the 1980s and began planting cocoa in 1989. In 1993 he switched to oil palm, as cocoa prices were too low and the crop had too many diseases. He planned to bring cattle onto the land but would need to build strong fences, as he already had 12 goats. He considered himself to be "pretty poor" (he claimed he received RM500 per month, while others with the same area of land earned RM1,800) and sent some members of his family away to work elsewhere. Like other farmers in the area, he hired Indonesian workers. In his case he had five Bugis workers.

Not far from this settlement there was a Bugis village (Kembara Jaya), whose members had managed to acquire land. The tenure situation of this group remains unsure, but in other areas Bugis rented land from the Orang Sungai. Their farming system was much more diverse, with large areas of lowland rice, some mixture of rice and young oil palm (which could work for the first two years, until the palm trees became too large), together with bananas, fruit and corn. Local people commented that these Indonesians worked very hard but were not necessarily permanent, with some going home while others took their place.

In another village (Sungai Lokan), people said they had arrived in 1985, with oil palm planting beginning two years later. The Department of Agriculture gave them assistance until the trees began bearing. As was the case with many farmers in this area, their families were large, with up to 11 or 12 children. Farmers here had 15 acres but complained that there was no more land for the children, who would have to go away. Although they said they did not earn very much from their own crop (pigs were a big problem, eating the fruit), they still preferred to be free from estate control. One farmer claimed he earned only RM250 per month, which was equivalent to estate wages of RM7 per day. An estate, established only in 1996, was very close to this village and had taken away some of the land that the people had believed was theirs, because of a technicality in the lease. They were now in a difficult position, without enough land to make a proper living: seven families had been affected. The estate was also accused of poisoning the water in the small tributary stream, making it impossible to fish there. In fact, there had been a flood, with all the pesticide from the estate lands finding its way into the river and killing several fish. Some families were now dependent on remittances sent by children working away in the cities. Wherever they could, the villagers also produced hill rice and a range of vegetables. One Indonesian was also found farming there, renting land from one of the villagers to grow chillies. Electricity was unreliable and services few, with most goods having to be purchased in Bukit Garam, which was also the location of the only secondary school. The Asiatic estate in Bukit Garam was also the receiving area for all the smallholders' oil palm, meaning that they had to pay extra transport costs. This village was certainly in an inferior position when compared with its wealthier neighbour, Bukit Garam, especially considering how scarce land was in the area, squeezed between oil palm estates.

In Bukit Garam, as elsewhere in Sabah, oil palm estate expansion has significantly reshaped the dynamics of land access and ownership, which in turn has redefined labour relations, migrations and access to capital. In that regard, oil palm plantations are by far the most important agent of territorial change on the east coast of Sabah.

Reshaping the Territory: For Whom?

The process of oil palm expansion is seen as the most fundamental process spearheading agrarian or territorial transition in Sabah. Two main arguments have been examined. (1) The colonial capitalist plantation system underpins the current plantation model, and as a result, (2) by the 1990s the post-independence agriculture expansion model of Malaysia, which had been based since independence on poverty alleviation, had been driven away from socio-economic development imperatives to serve capitalist expansion spearheaded by private corporations listed on the stock market. Although many in Sabah have also benefited from this form of development, this shift in the purpose of agricultural expansion in Sabah is reminiscent of colonial policy.

As this study demonstrates, the appropriation of large tracts of land by the corporate oil palm plantation sector has caused several real and potential problems for the rural populations in Sabah. These can be summarized as follows: Oil palm plantation expansion (1) deprives local populations of some of their basic livelihood conditions, such as a traditional pluri-activity system and territorial resources; (2) undermines the capacity of some communities to contribute to the conservation of biodiversity because of the limited areas available to villagers to improve their living conditions; (3) diminishes the capacity of some households to enhance their food security by growing other food crops because of land access problems; and (4) contributes to the erosion of "traditional territories", which may increase the vulnerability of populations that still partly rely on a wide range of non-timber forest resources. Although some smallholders are able to benefit from oil palm cultivation on their own plots, access to the benefits of oil palm plantation expansion is highly unequal.

The process of oil palm expansion affects an important number of people, whether Kadazan-Dusun, Murut or Orang Sungai. Migrant workers, who seem heavily dependent on land resources to grow food to complement their meagre wage from the estate, are also becoming more vulnerable as the expansion of oil palm estates continues.

From colonial to contemporary days, legislations and regulations pertaining to agricultural and territorial management have induced the economic integration of peripheral lands into the global market. Oil palm expansion in Sabah has become more tightly linked to globalization forces and global capitalist expansion, including the demand for new "green energy". Malaysian oil palm tycoons and their links with international capitalism currently constitute the main investors in agricultural expansion in Sabah, as well as in Southeast Asia as a whole (Casson 1999, Bernard 2006). Trade flows related to globalization have accelerated land-use conversion to commercial agriculture. However, the oil palm economy's configuration raises issues pertaining to its socio-economic contribution and environmental impacts, as the case of Kinabatangan District has illustrated. While territories with rich biophysical and cultural diversity are decreed empty for the purpose of oil palm development, the dominant oil palm plantation model creates enclaves largely detached from the socio-economic space in which it exists. This phenomenon is especially noticeable when the role played by local rural populations in the overall dynamic is examined. These populations, still characterized by high poverty rates, are largely kept outside of the oil palm plantation economy. When they are integrated through independent oil palm smallholdings linked to estate mills, for instance, the integration remains limited and occurs on highly uneven terms. Additionally, the small minority of populations still practising traditional swidden cultivation without formal land titles, especially in the interior of the state, endure the worst counter-effects of this relentless agricultural expansion.

The motives driving agricultural expansion in Sabah are in stark opposition to those behind the development programme embodied by FELDA. Although the FELDA model proved to be ill adapted to the realities of Sabah—providing a resettlement option to people who always had access to land—the current large-scale land conversion to oil palm plantations can hardly be legitimized by socio-economic development aims. Except for land taxes and sales taxes on palm oil, benefits derived from this activity are generally not reinvested locally, just like during the colonial era. Agricultural commercial expansion observed since the early 1990s, mostly in the form of large-scale estates, is a response to the primacy of local, national and global capitalist expansion but also to the maximization of the revenues of Sabah as well as Malaysia as a whole. Nevertheless, oil palm development has brought benefits to a number of smallholders. And without taking for granted that oil palm cultivation will continually expand in Sabah, it is quite likely that the growth of the sector will have a downstream industrial impact on employment in Sabah itself.

Historical institutionalization of natural resource exploitation and land commoditization currently facilitates oil palm plantation expansion in Sabah. Regulations enforced during the colonial days have favoured a rapid allocation of territories to this end, sparking a prompt response from corporations with readily available capital to supply the global market. The oil palm agricultural expansion model adheres to federal development policies whose prime objective, in accordance with the National Economic Policy, is to increase financial assets but without intervening in the redistribution of wealth. In addition, as with any other development project, the expansion of the oil palm cultivation domain is intrinsically political and can hardly be questioned in Malaysia. Nevertheless, as part of the government's modernization project, "Vision 2020", it legitimizes and precipitates profound territorial transformations for Sabah's inhabitants.

Notes

1. According to the Malaysian Palm Oil Board (MPOB 2008), 90 licences to produce agrofuel were allocated to various companies with a potential production capacity of 10.2 million tons a year. By May 2008, 12 plants were already operational with an annual potential production capacity of one million tons.

2. Because of its extent, the development of oil palm plantations raises numerous questions related to: (1) its impact on local populations and smallholders and their access to land (Cleary 1992; Beckford 2000; Doolittle 2001, 2003); (2) its consequences on the natural environment (Fitzherbert et al. 2008, McMorrow and Talip 2001, De Vantier et al. 2004, Hartemink 2005); and (3) economic strategies underlying this growth (Fold 2000, Sutton 2001).

3. According to Cleary (1992), quoting the *Handbook of the State of North Borneo, 1890*, the area officially allocated to European companies for tobacco cultivation reached 230,000 hectares. In 1910, 42,000 hectares were leased for rubber cultivation. However, because of high levels of land speculation (land leases were granted for 99 years), it is impossible to evaluate accurately the area actually devoted to crops. Given that land left idle during 12 to 20 years after the land lease was issued reverted automatically to the state, it can be assumed that most of this area had eventually been brought under cultivation.

4. The Sabah-based NGO Partners of Community Organisations (PACOS) defends native customary land's integrity through different forms of action such as community mapping. The organization supports native communities' land claims. The mandate of PACOS is to facilitate indigenous knowledge circulation and build exchange networks to improve natural resource management among native communities. This NGO, the most prominent in Sabah, favours cooperation with state agencies in order to provide services and advice to rural dwellers.

5. Contradicting previous adjudications, the Rumah Nor case extended to forested land and bodies of water the definition of customary land to which natives of Sarawak are entitled. The adoption of this definition of native customary land rights still depends on the judge's decision and his interpretation of the Land Code (Interview, Kota Kinabalu, July 2006).

6. On this matter, Doolittle (2001, 2004) highlights the stages of the evolution of native territoriality in colonial and postcolonial Sabah. According to her observations, individual landownership rights are progressively replacing community rights.

7. Interview with PACOS members, 12 July 2007.

8. Ibid.

9. Starting in the late 1950s, FELDA represented a powerful institution and a major actor in Malaysian economic development. According to guidelines of the New Economic Policy (NEP), development priorities of the Malaysian state focused on agricultural modernization, improvement of living standards and eradication of landlessness in rural areas in order to fight endemic rural poverty. Rather than attempting to redistribute existing wealth, the government decided to generate wealth through, in the case of FELDA, agricultural expansion (Bahrin and Lee 1988).

10. The companies mentioned merged at the end of 2006, creating the largest listed oil palm player on the global scene.

11. This concept identifies the "first" or allegedly "genuine" inhabitants of Malaysia: the Malays and the other native populations of Borneo. These populations were granted special economic rights in the wake of decolonization to catch up with more prosperous ethnic groups such as the Chinese. Since the 1969 ethnic riots in Kuala Lumpur between Chinese and Malays, the latter have enjoyed privileges in economy and education.

12. According to the PPB Group website, http://www.ppbgroup.com/, accessed Apr. 2007.

13. However, in the recent past federal authorities in Sabah have allegedly conferred citizenship on substantial numbers of immigrant populations from Indonesia and the Philippines who speak a language similar to the national official one and are easily integrated into the Malay-Muslim majority of the federation. Citizenship was granted to these migrant populations to inflate the ranks of ethnic Malays and therefore influence the results of subsequent elections. This strategy reinforced the electoral basis of the pro-federal party dominated by Malay-Muslims against autonomist factions in Sabah, dominated by autochthonous populations (Sadiq 2005).

14. The poverty referred to is calculated, according to the Eighth Malaysia Plan, by the establishment of a household budget threshold that guarantees access to sufficient daily calories, health care, clothing, transport, education and so on. According to recent adjustments in the computation of the poverty rate in 2005, this indicator for Sabah will probably be revised and figures scaled down.

15. It is, however, important to mention that the federal state has allocated increasing sums of investment for socio-economic development in Sabah over the last 15 years. This amounted to RM15 billion in the Eighth Malaysia Plan and represented almost 10 per cent of the federal budget for development in 2005. This increase in federal transfers for development in Sabah is intimately related to the evolution of political

relations between the two levels of government. The election of a pro-federalist government in Sabah has allowed the state government to break away from a confrontational anti-federalist approach. The end of disputes and sovereignty claims has put an end to federal budget cuts and retaliation measures, which had severe impacts on past budgets (Chin 1997, Moten 1999).

16. Dr Ongkili, Bernama, 13 July 2003. "Plantation Giants Should Plough Back Earnings – PBS", http://www.pbs-sabah.org.

17. The material in this section combines field notes compiled by Lesley Potter in 2005, field notes taken by Stéphane Bernard and Jean-François Bissonnette in 2007, as well as remote sensing and GIS calculations by Noé Pflieger, currently attached to the Lab THÉMA, Franche-Compté, France.

18. According to the Department of Agriculture, Sabah, the area under cocoa in 1990 was 205,976 hectares, while oil palm occupied 281,486 hectares. By 1998, the Sabah Institute of Development Studies reported "a gradual reduction in cocoa production attributed mainly to the increasing conversion of cocoa cultivated land into oil palm cultivation". Most cocoa is now confined to the Tawau area, farther south.

References

Appell, G.N., "Land Tenure and Development among the Rungus of Sabah", in _Modernization and the Emergence of a Landless Peasantry: Essays on the Integration of Peripheries to Socioeconomic Centers_, ed. G.N. Appel, Studies in Third World Societies Publication No. 33. Williamsburg: Department of Anthropology, College of William and Mary, 1985, pp. 111–58.

Bahrin, T.S. and Boon Thong Lee, _FELDA: Three Decades of Evolution_. Kuala Lumpur: FELDA, 1988.

Beckford, G.L., _Persistent Poverty: Underdevelopment in Plantation Economies of the Third World_. Jamaica: University of the West Indies Press, 2000.

Bernama, "Plantation Giants Should Plough Back Earnings, says PBS", http://www.accessmylibrary.com/coms2/summary_0286-23792449_ITM, accessed 22 Oct. 2010.

Bernard, S., _Palm Oil Expansion, Bio-fuel Production and Biodiversity Protection in Malaysia: Local Impacts of a World Global "Green Energy" Production Strategy_. Brisbane: International Geographical Union Conference, Queensland University of Technology, 2006.

Brookfield, H. and Y. Byron, "Deforestation and Timber Extraction in Borneo and the Malay Peninsula, the Record since 1965", _Global Environmental Change_ 1 (1990): 42–56.

Carter, C. et al., "Palm Oil Markets and Future Supply", _European Journal of Lipid Science and Technology_ 109 (2007): 307–14.

Casson, A., _The Hesitant Boom: Indonesia's Oil Palm Sub-Sector in an Era of Economic Crisis and Political Change_. Jakarta: CIFOR, 1999.

Chin, J., "Politics of Federal Intervention in Malaysia, with Reference in Sarawak, Sabah and Kelantan", _Journal of Commonwealth and Comparative Politics_ 35, no. 2 (1997): 96–120.

Cleary, M.C., "Plantation Agriculture and the Formulation of Native Land Rights in British North Borneo, 1880–1930", *The Geographical Journal* 158, no. 2 (1992): 170–81.

Courtenay, P.P., *Plantation Agriculture*. London: Bell Press, 1965.

De Vantier, A. Alcala and C. Wilkinson, "The Sulu-Sulawesi Sea: Environmental and Socioeconomic Status, Future Prognosis and Ameliorative Policy Options", *Ambio* 33, no. 1–2 (2004): 88–97.

Doolittle, A., "From Village Land to 'Native Reserve': Changes in Property Rights in Sabah, Malaysia, 1950–1996", *Human Ecology* 29, no. 1 (2001): 69–98.

————, "Colliding Discourses: Western Land Laws and Native Customary Rights in North Borneo, 1881–1918", *Journal of Southeast Asian Studies* 34, no. 1 (2003): 97–126.

————, "Powerful Persuasions: The Language of Property and Politics in Sabah, Malaysia (North Borneo), 1881–1996", *Modern Asian Studies* 38, no. 4 (2004): 821–50.

Ferguson, J., *Global Shadows: Africa in the Neoliberal World Order*. London: Duke University Press, 2006.

Fitzherbert, E.B., M.J. Struebig, A. Morel, F. Danielsen, C.A. Brühl, P.F. Donald and B. Phalan, "How Will Oil Palm Expansion Affect Biodiversity?" *Trends in Ecology and Evolution* 23, no. 10 (2008): 538–45.

Fold, N., "Oiling the Palms, Restructuring of Settlement Schemes in Malaysia and the New International Trade Regulations", *World Development* 28, no. 3 (2000): 473–86.

Goldthorpe, C.C., "A Definition and Typology of Plantation Agriculture", *Singapore Journal of Tropical Geography* 8, no. 1 (1987): 26–43.

Gomez, E.T. and K.S. Jomo, eds., *Malaysia's Political Economy: Politics, Patronage and Profits*. Cambridge: Cambridge University Press, 1997.

Hartemink, A.E., "Plantation Agriculture in the Tropics, Environmental Issues", *Outlook in Agriculture* 34, no. 1 (2005): 11–21.

Hayami, Y., "Family Farms and Plantations in Tropical Development", *Asian Development Review* 19, no. 2 (2002): 67–89.

Hugo, G., "Indonesian Labour Migration to Malaysia: Trends and Policy Implications", *Southeast Asian Journal of Social Science* 21, no. 1 (1993): 36–70.

IDS, *Institute for Development Studies, Sabah, Review of Sabah's Major Economic Indicators in 2004*, 2005, http://www.ids.org.my/current/indicators/.

Jomo, K.S., Y.T. Chang and K.J. Khoo, *Deforesting Malaysia: The Political Economy and Social Ecology of Agricultural Expansion and Commercial Logging*. London: Zed Books, 2004.

Juin, E., Y. Yangkat and C.H. Laugesen, "A Report on the State of the Environment in Sabah". Paper presented at the environmental convention in Kuching, Sarawak, 29–30 June 2000. State Environmental Conservation Department, Sabah, Malaysia.

Kummer, D.M. and B.L. Turner II, "The Human Causes of Deforestation in Southeast Asia: The Recurrent Pattern Is That of Large-scale Logging for Exports, Followed by Agricultural Expansion", *Bioscience* 44, no. 5 (1994): 323–8.

Li, T.M., "*Masyarakat Adat*, Difference, and the Limits of Recognition in Indonesia's Forest Zone", in *Race, Nature and the Politics of Difference*, ed. D.S. Moore *et al.* Durham and London: Duke University Press, 2003.

Lim, J.N.W. and I. Douglas, "The Impact of Cash Cropping on Shifting Cultivation in Sabah, Malaysia", *Asia Pacific Viewpoint* 39, no. 3 (1998): 315–26.

Liow, J., "Malaysia's Illegal Indonesian Migrant Labour Problem: In Search of Solutions", *Contemporary Southeast Asia* 25, no. 1 (2003): 44–64.

Long, B., J. Henriques, H.S. Anderson, Q. Gausset and K. Egay, "Land Tenure in Relation to Crocker Range National Park". ASEAN Review of Biodiversity and Environmental Conservation (ARBEC), Jan.–Mar. 2003, http://www.arbec.com. my/pdf/art5janmar03.pdf.

Majid-Cooke, F. "Vulnerability, Control and Oil Palm in Sarawak: Globalization and a New Era", *Development and Change* 33, no. 2 (2002): 189–211.

Malaysia, Government of, *Third National Agricultural Policy, 1998–2010* (NAP3). Kuala Lumpur: National Printing Department, 1998.

_____, *Sixth Malaysia Plan, 1991–1995*. Kuala Lumpur: National Printing Department, 1991.

_____, *Eighth Malaysia Plan, 2001–2005*. Kuala Lumpur: National Printing Department, 2001.

_____, *Mid-Term Review of the Eighth Malaysia Plan, 2001–2005*. Kuala Lumpur: National Printing Department, 2003.

Malaysian Palm Oil Board (MPOB), *Malaysian Oil Palm Statistics 2004*, http://www. mpob.gov.my/.

_____, *Malaysian Oil Palm Statistics 2008*, http://www.mpob.gov.my/.

McMorrow, J. and A.M. Talip, "Decline of Forest Area in Sabah, Malaysia: Relationship to State Policies, Land Code and Land Capability", *Global Environmental Change* 11, no. 3 (2001): 217–30.

Morgan, S., ed., *Ada Pryor: A Decade in Borneo*. London: Leicester University Press, 2001 [1893].

Moten, A.R., "The 1999 Sabah State Elections in Malaysia: The Coalition Continues", *Asian Survey* 39, no. 5 (Sept.–Oct. 1999): 792–807.

Ngidang, D., "Contradictions in Land Development Schemes: The Case of Joint Ventures in Sarawak, Malaysia", *Asia Pacific Viewpoint* 43, no. 2 (2002): 157–80.

_____, "Deconstruction and Reconstruction of Native Customary Land Tenure in Sarawak", *Southeast Asian Studies* 43, no. 1 (2005): 47–75.

Palm Oil Registration and Licensing Authority (PORLA), http://infolink.bernama. com/htmldocs/porla/porla.html, accessed 16 Dec. 2005.

Pang, T.W., "The Development Paradigm Shift in Sabah", in *Modern Malaysia in the Global Economy: Political and Social Change into the 21st Century*, ed. C. Barlow. Northampton: Edward Elgar Publishing, 2001, pp. 105–19.

Sabah, Government of, "State Budget Speech, Datuk Musa Haji Aman", 2006, http:// www.sabah.gov.my/, accessed 25 Feb. 2007.

_____, "Sabah Forest Plantation", www.sabah.gov.my/, accessed 2 May 2006.

_____, "Sabah Coastal Zone Profile", 1998, http://www.townplanning.sabah.gov.my/, accessed 20 May 2005.

_____, *Second Agricultural Policy, 1999–2010*. Kota Kinabalu: Ministry of Agriculture Development and Food Industry, 1999.

Sadiq, K., "When States Prefer Non-Citizens over Citizens: Conflict over Illegal Immigration into Malaysia", *International Studies Quarterly* 49, no. 1 (2005): 1001–122.

Sutton, K., "Agribusiness on a Grand Scale: FELDA's Sahabat Complex in East Malaysia", *Singapore Journal of Tropical Geography* 22, no. 1 (2001): 90–105.

Teoh, C.H., *Land Use and the Oil Palm Industry in Malaysia*. Abridged report produced for the WWF Forest Information System Database, 2000.

The Borneo Post, "Pairin Says There Is Need to Determine Actual Number of Illegals", 16 July 2006.

The Star, Malaysia, http://thestar.com.my/, accessed 24 Nov. 2006.

Thompson, S.J. and J.T. Cowan, "Globalizing Agro-Food Systems in Asia: An Introduction", *World Development* 28, no. 3 (2000): 401–8.

Wicke, B., V. Dornburg, M. Junginger and A. Faaij, "Different Palm Oil Production Systems for Energy Purposes and Their Greenhouse Gas Implications", *Biomass and Bioenergy* 32, no. 12 (2008): 1322–37.

Wood, E.M., *The Origin of Capitalism: A Longer View*. London and New York: Verso, 2002.

6

Agrarian Transitions in Kalimantan: Characteristics, Limitations and Accommodations

Lesley Potter
Australian National University

Introduction

The island of Borneo has been described by De Koninck (2006) as both "the ultimate frontier", with its low population density and abundant resources, and "the periphery of peripheries", a strategically situated land bank for capitalist development by Indonesia and Malaysia. Within the Indonesian two-thirds of the island, the four provinces of Kalimantan, one has heard over the past 15 years many grandiose plans and slogans for development. They began in 1995, with former President Suharto's "million-hectare rice scheme" in Central Kalimantan, a scheme that proved an abject failure and was a large contributor to greenhouse gas emissions during the fires of 1997. In 1998 the then Governor Suwarna of East Kalimantan proposed a million hectares of oil palm to form a "safety belt" in the northern border region. This million-hectare plan gradually moved south to include other sites in the province and was finally recognized as a failure in February 2005.[1] In May 2004 another million-hectare plan, again for oil palm, was announced for Central Kalimantan. A year later the central government announced that "the world's largest oil palm plantation" was to be established along the Indonesia-Malaysia border, to occupy 1 million hectares in West Kalimantan and 0.8 million hectares in East Kalimantan. That plan was revised downwards to 180,000 hectares in May 2006. Biodiesel was then acclaimed as the new engine of expansion, with the head of estate crops in South Kalimantan (a small province) suggesting, in October 2006, that there was potential there for 1.1 million hectares of oil palm to accommodate biodiesel. What is really happening in Kalimantan? Despite all the hype, there was a *total* area of only 1.1 million hectares of planted oil palm across the four provinces in 2005. This has expanded quite quickly since, to 1.66 million hectares in 2007 and 1.82 million hectares in 2008 (*Statistik Indonesia 2009*). Has Kalimantan already become largely commoditized? And what of local farmers and agricultural systems, both inside and outside the "oil palm belt"?

Agrarian Transitions: Some Theory and Kalimantan Examples

With the decline of employment in agriculture as a percentage of total employment and a slide in the proportion of GRDP attributable to agriculture, in Indonesia as elsewhere in Southeast Asia, it has become fashionable to conceptualize agrarian transitions, or "deagrarianization", as taking place throughout the region. More rural people are opting for non-farm work (even though continuing to live in villages) or moving to take up urban-based occupations (Rigg and Nattapoolwat 2001). The spread of oil palm plantations, now so characteristic of Kalimantan, does not quite fit this model, as it would appear to expand agricultural employment—with more intensive cultivation in a given land area—while at the same time introducing a new industry (the CPO factory) into the rural environment. The plantation is, however, a highly capitalized entity, producing largely for global export markets.

Those working in the industry are not peasants, in Elson's definition, as they have moved from partial dependence on subsistence production to reliance on a cash crop, and often need to buy their basic food (Elson 1997). Day labourers on the estates are entirely dependent on management decisions; attached smallholders have somewhat greater flexibility but are still constrained until they have repaid their establishment costs. Their ability to exercise agency is thus limited, though they certainly try. These workers, and surrounding local populations affected by the presence of the plantation, have also engaged actively in resistance to estate management decisions. They have used both the everyday forms of covert resistance described by Scott (1985, 1990) and more open shows of defiance such as demonstrations, road closures, destruction of seedlings, camp burning and seizures of machinery. Following the fall of Suharto in 1998 and a reduced role for the police and the army in enforcing estate management, overt resistance increased, but it has since declined (Potter 2009).

Jonathan Rigg (2005: 180) has proposed a general typology of agrarian transitions in Southeast Asia (Table 6.1), which is worth examining for its relevance to Kalimantan. The first major transition (which he suggests took place in the past) was the change from pure subsistence (with barter) to semi-

Table 6.1 Rigg's general typology of agrarian transitions in Southeast Asia

Type	*Agrarian type*	*Characteristics*
1	Subsistence	Farm- and village-focused; some barter and sale of surplus.
2	Semi-subsistence	Combine subsistence with commercially oriented agriculture; livelihoods remain farm- and village-focused.
3	Pluriactive (post-peasant)	Combine subsistence and commercially oriented farming with non-farm activities, both on- and off-farm. Migration and delocalisation of work are increasingly significant.
4	Professional	Professionalization of farming, emergence of agrarian entrepreneurs, larger-scale commercial enterprises; high levels of input; tightly integrated into national and international markets; technology-intensive.
5	Pluriactive (post-productive, neo-peasant)	Part-time farmers make lifestyle choices, combine farming with other occupations, trading higher income for better quality of life.
6	Remnant smallholder	Rural households who remain tied to the land and to traditional production systems. Productivity is low, poverty high.

Source: Summarized from Rigg 2005.

subsistence agriculture (with some commercial cropping). Present transitions include, most notably, the move from semi-subsistence to "pluriactive" (with increasing involvement in non-farm activities and migration) and from "pluriactive" to "professional", by which Rigg means large-scale, high-input farming, integrated into national and international markets. It is possible that the fourth transition, to a kind of hobby farming, more common in developed countries, might eventually occur in the future, but it has no present role in Kalimantan. Rigg's final category, of "remnant smallholder", might be found on the fringes of modern development: while there is no transition that specifically produces such a type, it is easy to imagine such households existing in marginal locations.

In addition to Rigg's categories, use will be made of the organizing framework of De Koninck (2004), who identified six main features of the modern agrarian transition in Southeast Asia: agricultural intensification; increasing integration of production into the market; higher levels of industrialization and urbanization; greater population mobility; new forms of regulation; and a revaluation of environmental resources. It is clear that the two sets of ideas are similar, though some details differ. This paper will analyze the extent to which aspects of both Rigg's transitions and De Koninck's framework may be applied to the four provinces of Kalimantan. There will be a concentration on the question of agricultural intensification, through the partial replacement of swidden farming by plantation oil palm as production becomes increasingly market-oriented. Other aspects, such as comparative levels of agricultural versus urban-based employment, types of industrial and mining activity, population mobility and regulation will also be examined.

Rigg's Transitions Examined: The Roles of Rubber, Multiple Activities and Plantation Oil Palm

Rubber: Transition to Semi-subsistence Agriculture

While there is room for discussion on whether subsistence systems in Borneo were ever completely inward-looking or reliant on barter, given the centuries-old trade in collected forest products,[2] the first agrarian transition, from subsistence to semi-subsistence, is hypothesized to have occurred in the first three decades of the 20th century following the introduction of rubber, the South American exotic *Hevea brasiliensis*. Rubber became integrated into existing food-crop systems as the most important income source for small farmers (Dove 1993). It could be argued that the existence of rubber gardens caused some land to be more intensively farmed, though it did not eliminate swiddening among traditional Dayak groups. Its effect was to move traditional systems away from pioneer swiddens with nomadic settlements to rotational swiddening closer to permanent villages. More productive "wet swiddens" (*padi paya*), a precursor to the complex water control of wet rice production (*padi sawah*), also gradually became popular where water conditions permitted. A further transition, from *paya* to *sawah*, has also been taking place, as noted by Padoch *et al.* (1998).

The role of rubber as a cash crop has differed across Kalimantan. It has not been very important in East Kalimantan,[3] where pepper and rattan were alternatives but smallholder cash cropping was never as prominent an income source as forest work or mining. Rubber did not penetrate into the most remote swidden-based villages of West Kalimantan or Central Kalimantan until the 1930s. However, it was planted by Chinese farmers in coastal West Kalimantan in 1903, with the cultivation quickly spreading up the Kapuas River to the Dayak lands of Sanggau, where its development was facilitated by Chinese merchants. By the time of high rubber prices in 1925, Dayaks were reported to be giving up their nomadic life and becoming more attached to the soil (Loos and Van Beusechem 1925), with rubber well incorporated into traditional systems (Ozinga 1940). The crop was also well established by 1925 around the edges of the wet rice bowl of the Hulu Sungai region of South Kalimantan. Banjarese farmers, considered by Dutch observers as the most commercially enterprising of the Malays, were firmly focused on the international market, with Singapore prices quoted by merchants in the busy *pasar* at Kelua near Tanjung. A well-developed road system meant that motor vehicles as well as bicycles and ox-carts played a role in bringing the rubber to the regional markets, though the traded product was still shipped downriver to Banjarmasin (Luytjes 1925).[4] Although a few large rubber plantations were established in both West Kalimantan and South Kalimantan, the industry was and has remained overwhelmingly smallholder-based.

While rubber is a flexible crop, needing little local processing and being able to grow successfully in a kind of secondary forest with other tree species (where it is known as "jungle rubber"), it has suffered from both extreme price variations and low productivity, especially when trees become senescent. The central government tried for decades to improve stock by means of block planting schemes,[5] but access to improved seeds has not been easy for smallholders, most of whom have remained reliant on traditional varieties. In June 2007 it was estimated that there was an annual shortage of 20 million improved seeds, demand for which was high as rubber prices were rising steadily (*Kompas*, 6 June 2007). Programmes promised by the central government for "revitalizing" rubber production offer financial assistance to growers to buy seed and fertilizer, but amounts are still considered inadequate. West Kalimantan, the leader of the four provinces in area under rubber (538,000 hectares in 2007), had ambitious plans to more than double its cultivation area to 1.2 million hectares by 2009 but would need large additional funding, especially to replant an estimated 88,000 hectares of senescent trees (*Pontianak Post*, 28 Feb. 2007). Border areas with Sarawak are especially targeted: villagers in those regions had lived mainly on proceeds of the timber industry from 1999 to 2006, but the push to eliminate illegal logging was forcing them back to a reliance on farming, particularly rubber (*Pontianak Post*, 30 Dec. 2006).

"Pluriactivity" and Migration

Rigg's "pluriactivity" type of agrarian system raises some questions as to its recency as a phenomenon. Many peasant households have long been engaged in a range of occupations that supplemented basic rice production and reduced risk. These have included small-scale mining; collecting and selling forest products; making roof thatch, shingles, mats and other handicrafts; boatbuilding; fishing; and animal raising.[6] However, Rigg distinguishes the plural activities in this transition as "post-peasant", for example, industrial developments extending into villages, as he and Nattapoolwat discovered in Thailand (Rigg and Nattapoolwat 2001), and the migration (sometimes circular) to work in urban centres. In South Kalimantan Banjarese have moved from the Hulu Sungai to work in industries in Banjarmasin, but rural-to-rural seasonal migrations have also been common, such as the movement from one district rice harvest to another or adventitious dry season cropping in peat swamps. Travel from the Hulu Sungai to the harvest at Gambut, near Banjarmasin, would take place each August. In 1991 this attracted 50,000 participants, the majority of whom were women (Potter 1993). These alternative income sources are also an indication of poverty in the home districts. Lack of sufficient land (an average of only 0.5 hectare of *sawah* land per family) and the fact that much of the existing land is owned by urban-based absentees have been serious problems in the Hulu Sungai and are certainly drivers of the search for alternative, non-rural occupations (*Banjarmasin Post*, 8 May 2002). This particular problem is unique to this part of Kalimantan and is more akin to the situation in Java.

The Transition to "Professional" Farming

Most compelling of the recent agrarian transitions taking place in Kalimantan has been the growth of large-scale plantations, especially of oil palm, which fits Rigg's category of "professional" even though it is not local farmers but outside corporations who own the estates. The oil palm plantations score well on the other descriptors, such as market integration and high capital input. Although local entrepreneurs are occasionally found in the agricultural sector, actual ownership is more often Jakarta-based, Malaysian or Singaporean. This chapter is largely concerned with this particular transition and attempts to assess its extent, the types of people affected and the position of those outside the "oil palm belt".[7] Do the latter households fit Rigg's final category of poverty-stricken "remnant smallholders" or are they involved in alternative activities?

Further Questions of Population and Mobility

Another aspect of the "agrarian transition" model as suggested by both Rigg and De Koninck is increased population mobility. In Kalimantan two ethnic groups are well known for their mobility: the Banjarese of South Kalimantan and the Iban. The Iban are found mainly in the border regions of West

Kalimantan, with larger Iban populations in Sarawak (East Malaysia). Sarawak is now the focus of many Indonesian Iban, as they move easily to work across the porous border. Other Dayak groups were also mobile from the late 19th century, with Kenyah and Kayan in East Kalimantan moving away from their original mountain homelands towards the coast and potential income sources, a movement later continued and encouraged by the Suharto government.

Two hundred kilometres inland from Banjarmasin, the heartland of Banjarese culture, the productive alluvial fans of the Hulu Sungai were considered *overpopulated* in 1893, even though population density in the rest of Kalimantan was very low (Joekes 1894). Outmigration of Banjarese began in the 1860s and has continued to modern times. They moved to parts of present East Kalimantan and Central Kalimantan (where they form an important segment of the ethnic mix), later to Sumatra and Peninsular Malaysia and then to urban Banjarmasin (Potter 1993). The Hulu Sungai, a famous centre of wet rice production, with rubber along its hilly eastern and northern margins, continues to have higher rural densities and much lower rates of population growth than any other part of Kalimantan (Figure 6.1).

Kalimantan's east coast has long been a migration destination for peoples from Sulawesi—especially Bugis, but also Mandar and Toraja. Originally wet rice farmers and fishers, then pepper and cocoa growers, they now form an important component of the modern mining and industrial workforce.

The impact of mobility may be seen in another way, due to the influx of transmigrants into what were perceived by government authorities as "empty" areas. In Central Kalimantan, 65 per cent of the population increase between 1980 and 1985 resulted from transmigration; comparable figures for West Kalimantan and South Kalimantan were 46 per cent and 41 per cent respectively (Potter 1991). East Kalimantan was less reliant on government-sponsored movement, as mining, industrial and forestry activities readily attracted a spontaneous inflow.[8] Government-sponsored transmigrants (primarily from overcrowded Java) had no choice in their location. Many schemes were unsuccessful in their aim to establish Javanese wet rice growers and wean locals away from swidden agriculture. In fact, the reverse often occurred, with desperate Javanese learning survival skills from more knowledgeable indigenous farmers (Hidayati 1994).

Transmigrants were also used as a labour force to support tree crop plantations, initially for estates growing improved varieties of rubber, then more and more for oil palm. The World Bank and the Asian Development Bank encouraged a joint "nucleus and smallholder" type of development, initially on government-owned estates, with smallholders (*plasma*) receiving 2–3 hectares of oil palm but also working as labourers on the estate nucleus. A World Bank

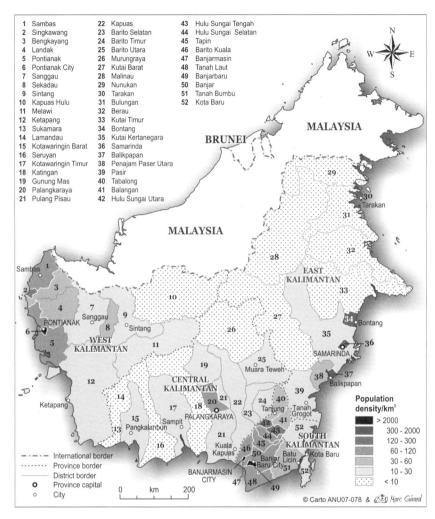

1	Sambas	22	Kapuas
2	Singkawang	23	Barito Selatan
3	Bengkayang	24	Barito Timur
4	Landak	25	Barito Utara
5	Pontianak	26	Murungraya
6	Pontianak City	27	Kutai Barat
7	Sanggau	28	Malinau
8	Sekadau	29	Nunukan
9	Sintang	30	Tarakan
10	Kapuas Hulu	31	Bulungan
11	Melawi	32	Berau
12	Ketapang	33	Kutai Timur
13	Sukamara	34	Bontang
14	Lamandau	35	Kutai Kertanegara
15	Kotawaringin Barat	36	Samarinda
16	Seruyan	37	Balikpapan
17	Kotawaringin Timur	38	Penajam Paser Utara
18	Katingan	39	Pasir
19	Gunung Mas	40	Tabalong
20	Palangkaraya	41	Balangan
21	Pulang Pisau	42	Hulu Sungai Utara
43	Hulu Sungai Tengah		
44	Hulu Sungai Selatan		
45	Tapin		
46	Barito Kuala		
47	Banjarmasin		
48	Tanah Laut		
49	Banjarbaru		
50	Banjar		
51	Tanah Bumbu		
52	Kota Baru		

Figure 6.1 Kalimantan. Population density by district, 2000

report (1989) recommended that more private estates be involved, anticipating that most of the labour force would be transmigrant. The rapid expansion of oil palm cultivation during the 1990s was partly due to the application of that model throughout Kalimantan, with the proliferation of private, rather than government, estates, the latter now forming only a small segment of the whole. When only transmigrants are given access to oil palm smallholdings, local people feel marginalized and may resort to violence to secure their participation. All new estates must now include a local smallholder component, though the

Table 6.2 Kalimantan. Major ethnic groups (%) by province, 2000

Kalimantan	South	Central	West	East
Javanese/Sundanese	13.8	17.8	9.4	31.1
Banjarese	75.6	19.9	2.0	13.9
Bugis/Toraja/Mandar	3.8	…	3.3	20.2
Madura	1.6	2.0	5.5	…
Chinese	…	…	10.0	…
Dayak	1.9[c]	60.0[ad]	33.8[e]	34.0[ab]
Melayu	…	…	33.8	…
Total	96.7	99.7	97.8	99.2

Notes: a: Estimate; b: Includes Kutai, Paser, Kenyah, Kayan, Tidung, Punan, etc.; c: Includes Bakumpai, Buket, Bukat, Dusun, etc.; d: Includes Ngaju, Maanyan, Bakumpai, Sampit, Katingan, Ot Danum, Lawangan, Tamuan, etc.; e: Includes Iban, Hibun, Desa, Ketungau, Jelai, Persaguan, Jagoi, Maloh, Bukat, Kenyilu, Punan, Taman, etc.

Sources: BPS 2001: Kalimantan Selatan, Kalimantan Tengah, Kalimantan Timur; BPS Pontianak 2003: Kalimantan Barat.

nature of this has been changing. The transmigration programme was wound back after the fall of Suharto in 1998 and decentralization in 2001.[9]

The overall impact of all this migration has been to produce a multi-ethnic population in much of Kalimantan, with Javanese transmigrants and Bugis and Banjarese spontaneous migrants, various Dayak groups (often in the interior), and indigenous Malays (often near the coast). Descendants of Chinese gold miners form a specific minority in West Kalimantan, which is now largely urban, especially in Pontianak and Singkawang (Table 6.2).

The final aspect of mobility that is currently important is international migration. In Kalimantan, most international movement is north towards the East Malaysian states of Sarawak and Sabah, though a trickle of people, mainly women, has gone to Saudi Arabia. The long border between East Kalimantan and West Kalimantan and Sarawak has seen a considerable movement of workers (many of them illegal), from Sambas (West Kalimantan) to the oil palm estates of Miri or from Kerayan in East Kalimantan to the rice fields of the Kelabit Highlands. On the other hand, the workers moving into Sabah from Nunukan (the northernmost district of East Kalimantan) have been largely Bugis from Sulawesi or Floreanese from Nusa Tenggara Timur seeking employment in the oil palm or timber industry. Some female Dayaks are employed as domestics in Sarawak, while there is considerable illegal trafficking of women and girls into Malaysia.

What does this increased mobility actually mean in terms of agrarian transition theory? While some movements have been noted as directed towards cities and non-rural occupations, opportunities for unskilled local migrants in Kalimantan (and in East Malaysia) are often still in rural areas. The transmigration projects have been specifically directed to rural locations, especially to provide part of the palm oil workforce. While East Kalimantan, with its high levels of urbanization, is different from the other provinces, its cities do not attract many local Dayaks but are home to a varied mix of migrants from many parts of Indonesia.

The Structure of Employment and Provincial Economies, 2006–7

Bearing in mind the above caveats, but in line with agrarian transition findings in other countries and seemingly supporting De Koninck's framework, there has been a reduction in levels of employment in the primary industries (agriculture, forestry, hunting and fishing), concomitant with higher levels of involvement in manufacturing, trade and services, usually urban based. This trend is especially noticeable in East Kalimantan, which, together with South Kalimantan, also has strong mining industries based on petroleum and coal. Table 6.3 presents

Table 6.3. Kalimantan. Employed population 15 and over by province and main industry (2007) and projected levels of urbanization by province (2005 and 2010)

Kalimantan	*South*	*Central*	*West*	*East*	*Average*
Agriculture, forestry, fishing	45.30	59.2	62.5	33.9	51.6
Mining	3.50	4.4	2.2	5.7	3.6
Manufacturing	8.20	4.3	4.3	7.6	6.0
Utilities	0.04	0.1	0.1	0.4	0.2
Construction	3.60	3.9	3.7	6.3	4.2
Transport, communications	4.90	4.2	3.1	6.8	4.5
Financial services	0.90	0.8	0.6	2.4	1.1
Public services	12.70	9.5	9.5	15.6	11.6
Total	79.14	86.4	86.0	78.7	82.8
Population, % urban 2005 (projected)	41.50	34.0	27.8	62.2	41.3
Population, % urban 2010 (projected)	46.70	40.7	31.1	66.2	45.8

Sources: *Statistik Indonesia* 2008, Table 3.2.4; Proyeksi Penduduk 2000–2025 - 3.5: Urbanisasi.

figures on employment by major industry, together with projected levels of urbanization.[10] While the employment figures are summary by nature and do not take account of the "pluriactivity" emphasized by Rigg, they nevertheless reveal clear differences between East Kalimantan and the more agrarian provinces of West Kalimantan and Central Kalimantan, so it is perhaps not surprising that the latter have a larger area planted with oil palm.

The trend is even more marked when one examines the structure of the provincial economies (Table 6.4), even though East Kalimantan's percentages are distorted by the value of the oil and gas sector and are also shown with that sector excluded. East Kalimantan is one of Indonesia's wealthiest provinces, but much of the wealth is enclave in nature, with manufacturing industries (especially LNG and fertilizer, located in the city of Bontang) heavily dependent on oil and gas. When that sector is excluded, the value of coal mining is emphasized: agriculture and forestry still do not rate highly in terms of value. While industries in the major cities of Samarinda, Banjarmasin and Pontianak have in the past been heavily involved in sawmilling, plywood manufacture and other types of wood processing, these have been declining in recent years, especially in Samarinda. Such industries were hit hard by the

Table 6.4. Kalimantan. Structure of gross regional domestic product at current prices (%) by province and total value of product (million Rp), 2006

Activity	South	Central	West	East plus oil & gas	East minus oil & gas
Agriculture, forestry, fishing	22.8	35.4	27.0	5.3	13.1
Mining	21.3	6.8	1.2	41.6	33.4
Manufacturing	12.1	8.5	18.5	36.2	12.1
Utilities	0.5	0.5	0.6	0.3	0.7
Construction	6.4	5.0	8.6	2.4	5.8
Trade, hotels, etc.	14.7	18.1	22.7	6.5	15.9
Transport, communications	8.6	9.8	6.8	3.5	8.6
Financial services	4.1	4.6	5.2	1.7	4.3
Public services	9.5	11.3	9.4	2.5	6.1
Total	100.0	100.0	100.0	100.0	100.0
Total value (million Rp)	34,469,335	24,401,554	37,714,997	198,579,232	80,964,021

Sources: Kalimantan Barat Dalam Angka 2007, Kalimantan Selatan Dalam Angka 2007, Kalimantan Tengah Dalam Angka 2007, Kalimantan Timur Dalam Angka 2007.

global financial crisis of 2008–9 as orders diminished, while raw materials became hard to find (*Tribun Kaltim*, 28 Apr. 2009). It is worth noting that in 1975, agriculture occupied more than 50 per cent of the value of the GRDP in both West Kalimantan and Central Kalimantan but only 13.4 per cent in East Kalimantan; thus, the differences between the provinces are of long standing (Vidyattama *et al.* 2007).

Kalimantan and Oil Palm: Background and Overview

Some Historical Background

The African oil palm (*Elaeis guineensis*) arrived at the Botanic Gardens in Bogor (West Java) in 1848. A few seedlings were transferred in 1875 to the "plantation belt" of northern Sumatra, where the industry was eventually established in 1914, with the first exports in 1919. Though the crop was very successful, making Holland a world leader in palm oil production by the 1930s (Blommendaal 1937, Thee 1977), it remained confined to parts of North Sumatra and Aceh. No tradition of smallholder oil palm cultivation developed on or near those estates.

In 1975, 100 years after the beginnings of oil palm cultivation in Sumatra, the governor of West Kalimantan suggested the introduction of oil palm plantations into his province, "to utilize the critical lands" (PNP VII 1984: 15). "Critical lands" were generally swidden fallows carrying a vegetation of secondary regrowth, scrub and *Imperata* grassland. They were also called "sleeping land" (*lahan tidur*) by Javanese government officials. By targeting the swidden fallows, the governor was seeking to force intensification of Dayak agriculture through limiting its land base, bringing "development" in the form of plantations to what was considered a backward province.[11] Despite the growth of rubber as a cash crop, local farming was described as still in a semi-subsistence condition by Australian consultants:

> Land is used under traditional forms of tenure, labour is mobilised on a kinship basis, trade and exchange are often non-monetary, and the social system is resistant to many aspects of technical and economic innovation (Ward and Ward 1974: 53).

A survey was carried out in 1978 for the government plantation company PNP VII of North Sumatra, the decision being to set up estates in Sanggau District and at Ngabang, closer to Pontianak. In 1984, after many early difficulties and cultural misunderstandings, there were 5,000 hectares of oil palm in West Kalimantan, employing predominantly local Dayak labourers and smallholders (Potter 2009). The same year another government-owned plantation, PTP VI, began operating at Long Ikis in Pasir District, East Kalimantan, using Javanese transmigrant labour (Mayer 1988). South Kalimantan's first oil palm estate (privately owned by Sumatran interests) was established in Kota Baru on the east coast in 1981 but had only 440 hectares planted when the author visited in October 1986. As in South Kalimantan, there were no government oil palm estates in Central Kalimantan. The first two private oil palm plantations were set up in 1992, one in Kotawaringin Barat, north of Pangkalanbun, near the boundary between the mineral soil and the peat swamp forest, the other far up the Barito River near the town of Muara Tewe (Kalimantan Tengah 2009). The industry was gradually established as the Trans-Kalimantan Highway was constructed between Pangkalanbun and Sampit, with more plantations rapidly following along that route. It was only in the 1990s that the industry began its rapid expansion into all four provinces of Kalimantan.

Palm oil is not as easy for smallholders to produce as rubber, because of the need for ripe oil palm fruit to be processed within 48 hours of picking. As factories are expensive to construct, this processing is generally done by a well-capitalized plantation enterprise, so that smallholders are linked to a plantation, which restricts their freedom of choice. The province with the most "free" smallholders is Riau (Sumatra), where "mini" factories have been built, often through a cooperative to serve their needs. Such mini factories have not yet appeared in Kalimantan, though suggestions for their development have been made from time to time.

General Overview: Physical and Cultural Constraints on Oil Palm Expansion

A striking feature of Kalimantan's physical structure is its high central mountain and plateau area, with individual peaks exceeding 2,000 metres, especially in East Kalimantan, where it attains its broadest extent. Much of this elevated area is included in the national parks of Kayan Mentarang (East Kalimantan) and Betung Kerihun (West Kalimantan), which form part of the international "Heart of Borneo" conservation initiative, together with Bukit Baka/Raya National Park and the proposed Muller Mountains National Park, both in Central Kalimantan. A smaller range (the Meratus Mountains) extends north-south through South Kalimantan. The interior highlands are the source of the major streams flowing west (Kapuas), south (Barito, Katingan, Kahayan, Seruyan and others) and east (Mahakam, Kayan, Mentarang and others). These streams once provided the sole means of transportation for human activities, and although they retain some importance they have been partly replaced by roads. The coastal areas are often swampy for a considerable distance inland, with varying thicknesses of peat (especially in the west and south). Neither the highland areas nor the peat swamps are really suitable for oil palm plantations, though there have been some recent incursions into the peat areas, especially in Central Kalimantan (Hooijer *et al.* 2006). Between the swamps and the mountains lies an area of flat to undulating land, once completely covered by lowland forests. Considerable forest remains in East Kalimantan and Central Kalimantan, and where accessible it has been targeted for clearing, ostensibly for plantation development. The most important sites for oil palm generally

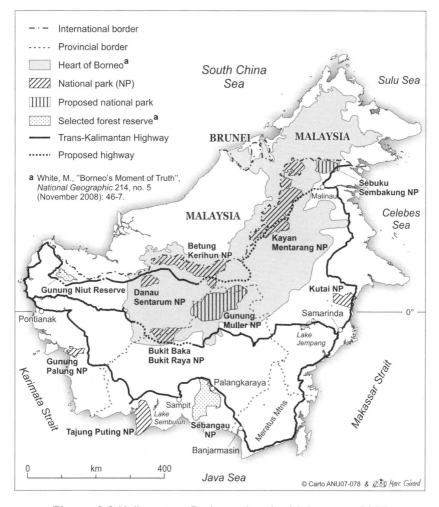

Figure 6.2 Kalimantan. Parks and major highways, ~2007

lie within reach of existing road transport facilities. The Trans-Kalimantan Highway, despite its poor condition for much of its length, remains a concentration point for oil palm plantations (Figure 6.2). There are significant gaps in this road, especially in its north link between Putussibau and Malinau, where it must cross high mountains.

Using 2004 or 2005 data, a summary can be established, at sub-district level, of the proportion of land that was occupied by planted oil palm (Figures 6.3 to 6.6). As each oil palm estate controls more land than it has actually planted, the figures underestimate the impact of oil palm on the lands of in situ populations—but they do accurately show the major regions in which the crop

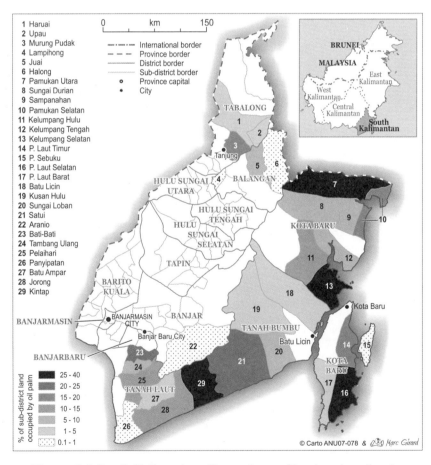

Figure 6.3 South Kalimantan. Percentage of land under oil palm cultivation, by sub-district, ~2005

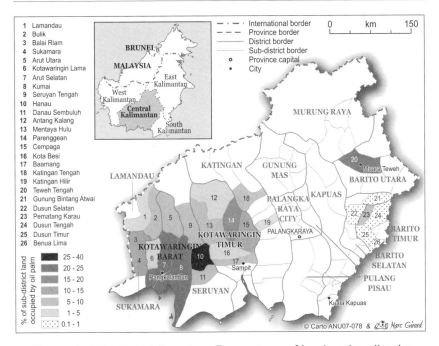

Figure 6.4 Central Kalimantan. Percentage of land under oil palm cultivation, by sub-district, ~2005

is concentrated. The effect produced by these choropleth maps is also a factor of the size of sub-districts, which tend to be larger in East Kalimantan (Figure 6.6), so provide a less fine-grained mosaic of actual estate locations.[12] Oil palm holdings are scattered across each of these sub-districts but do not reach to the swampy coasts, as they appear to do in the districts of Kotawaringin Barat (Central Kalimantan, Figure 6.4) and Pontianak (West Kalimantan, Figure 6.5). It is clear from the maps, however, that oil palm is clustered in specific parts of all four provinces, so one may see an oil palm belt and a quite extensive area where oil palm has not been planted. Within parts of the latter, location permits may have been taken out and forests cleared, but no further activity had ensued at the time the data were collected.[13]

Environmental limitations may partly explain this clustering, but social factors are also important in influencing the rate of oil palm development. West Kalimantan, the first province to experience the loss of traditional Dayak land to plantation companies, has developed a strong NGO lobby that is actively discouraging further allocation of land to oil palm. NGOs in West Kalimantan support rubber, pepper and cocoa as alternatives. Active resistance, which has

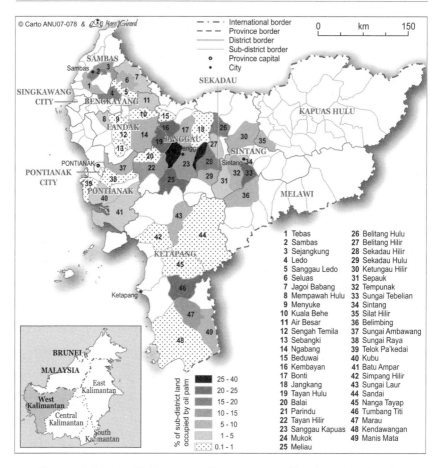

Figure 6.5 West Kalimantan. Percentage of land under oil palm
cultivation, by sub-district, ~2005

occurred spasmodically throughout Kalimantan's oil palm belt, was highest
soon after the fall of the authoritarian Suharto government in 1998, and lasted
through the transitional reformasi period to the beginning of decentralization
in 2001. At that time commodity prices for both rubber and oil palm were
very low. Subsequent increases in price have helped dampen resistance, though
violent confrontations still occasionally occur. District authorities are invariably
supportive of oil palm companies and have recently become less tolerant of
violent reactions from local communities, being ready to call on the police to
restore order, an activity reminiscent of the Suharto period. New companies
usually employ local officials to "socialize" villagers to the benefits of signing

over some or all of their lands in return for 2 hectares of oil palm. It is at this point that NGOs are beginning to step in to counter company "propaganda" and to provide alternative information, especially informing villagers of the large interest bill they will have to pay to cover the cost of establishing their smallholding. The fact that they will have to surrender their freedom of decision over their cultivation for at least eight to ten years, while the company or the cooperative has control, often makes villagers pause and perhaps reject the company's advances.

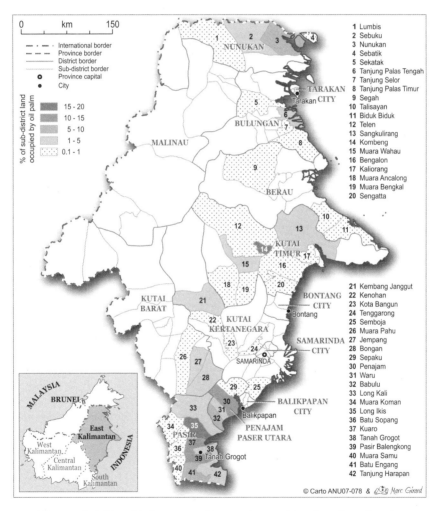

Figure 6.6 East Kalimantan. Percentage of land under oil palm cultivation, by sub-district, ~2005

Land tenure has been a crucial factor in both the spread of the crop and local resistance. There may still be "empty" areas in Kalimantan, but most land, even *Imperata* grassland, is under traditional claims, though these may be legally weak. All lands classified as "forests" and not in permanent agriculture remain the property of the state, which authorizes their leasing to logging and plantation companies and, under the revised Forestry Law (No. 41 of 1999), to smallholder cooperatives. According to the 1980s forest classification (Tata Guna Hutan Kesepakatan), agricultural activities, including the cultivation of estate crops such as oil palm, may take place only in areas of "conversion forest".[14] "Production forest" is supposed to be occupied by logging concessions or tree plantations developed for reforestation or pulp. The boundary between production and conversion forest has always been contested: as forests have been logged, pressure has grown for their reclassification. Oil palm companies wishing to access "production forest" had to apply to the Minister of Forestry to have the area excised, and show proof that the forest on the land was degraded. This process could take years, so local authorities often provided bridging authorization for potential investors (Potter and Lee 1998). Since decentralization in 2001, the maintenance of central authority over lands and forests has been disputed and some district leaders have attempted to override it, but legally the Minister of Forestry still holds power. In April 2007 it was announced by the minister that no more conversions from production forest to agriculture would be allowed. The ministry stated that 23 million hectares across Indonesia had been converted but only 2 million had resulted in planted oil palm: the rest had been neglected. New plantings would have to use that neglected land (*Kompas*, 10 Apr. 2007).[15]

During the Suharto regime, companies would simply take over village land, felling fruit trees, rubber gardens, traditional forests and forested fallows without permission, sometimes without warning and generally with little compensation. Village heads might be bribed into handing over communally owned lands, or tricked with inaccurate maps. Where necessary the police and the army would be called in to ensure compliance, though it was possible to opt out and to retain one's land as an "enclave" inside the plantation. Depending on the type of plantation envisaged, villagers were sometimes offered participation in a scheme, but they would always have to give up more land than they actually received, to provide for the nucleus of the estate and perhaps transmigrant smallholders as well.

The current Yudhoyono government (especially Agriculture Minister Anton Apriyantono and Director of Estate Crops Achmad Manggabarani) has continued earlier policies presenting palm oil as the "prima donna" of agricultural exports, while Indonesia has displaced Malaysia as the world's leading producer, though still lagging slightly in levels of export (Oil World Annual 2009). Plantation companies have been encouraged to shift their focus from their traditional bases in Sumatra to Kalimantan, Sulawesi and Papua, but this has been only marginally successful.[16] There are certainly Sumatra-based plantations that have expanded into Kalimantan, but the comparatively poor infrastructure of the latter has been a deterrent. The heads of new districts formed since 1999 as part of Indonesia's decentralization process have been especially keen to encourage investment in oil palm, as under Law 25 of 1999 (Fiscal Balancing) they must raise significant income from local sources. They rely on the big companies to increase district tax revenues and provide employment, and they tend to brush aside complaints of undesirable company practice from villagers or NGOs. The very high prices for crude palm oil (CPO) experienced during 2007 and the first half of 2008, largely due to the international demand for biodiesel from both Europe and China, initiated a new frenzy for the development of oil palm across Kalimantan, with many districts ignoring the environmental and social consequences. However, with the drop in petroleum prices resulting from the world economic crisis, demand for biodiesel has declined and CPO prices have plummeted. Prices received by farmers for their fresh fruit bunches dropped below the cost of production during the worst months of October–December 2008, while the factories were processing only the best-quality fruit.[17]

A comparison of the growth of the commodity in the four Kalimantan provinces from 2001 to 2007 is revealing (Table 6.5). The slower growth rates in West Kalimantan reflect the more difficult social situation there and the greater success of NGOs in promoting resistance. The older age of many trees has also produced a crisis in replanting, as yields have dropped. West Kalimantan, after historically being the leader in area planted with oil palm, had to give way to Central Kalimantan in 2004, and the latter has continued to power ahead. Despite many difficulties and the "false starts" associated with bogus companies interested only in accessing timber, East Kalimantan achieved the highest overall rates of progress in establishing oil palm, with a huge increase in 2006–7. This province still lags well behind Central Kalimantan and West Kalimantan, however, partly because of its industrial base.

Oil Palm and Its Alternatives in the Four Provinces

The four provinces of Kalimantan are very different in their background resources, economic level and present adaptation to oil palm. This section of the paper will outline these differences, followed by a description of the impact of oil palm in the most important locations (Figures 6.3 to 6.6), and the

Table 6.5 Kalimantan. Oil palm-planted area by province and rate of growth,
2001–7

Province	South	Central	West	East	Total
2001	137,137	252,625	323,080	117,055	829,897
2002	153,745	295,946	335,896	132,174	917,761
2003	155,668	343,323	349,101	159,079	1,006,878
2004	172,650	401,662	367,619	171,581	1,113,512
2005	173,392	405,554	381,566	201,087	1,161,599
2006	243,451	571,874	407,083	225,337	1,447,745
2007	253,984	616,331	451,400	339,292	1,661,007
Rate of growth 2001–2002 (%)	12.1	17.1	4.0	12.9	10.6
Rate of growth 2002–2003 (%)	1.2	16.0	3.9	20.4	9.7
Rate of growth 2003–2004 (%)	10.9	17.0	5.3	7.9	10.6
Rate of growth 2004–2005 (%)	0.4	1.0	3.8	17.2	4.3
Rate of growth 2005–2006 (%)	40.4	41.0	6.7	12.1	24.6
Rate of growth 2006–2007 (%)	4.3	7.8	10.9	50.6	14.7

Sources: Perkebunan Dalam Angka, Kalimantan Barat 2008, Kalimantan Selatan
2008, Kalimantan Tengah 2008, Kalimantan Timur 2008.

position of more remote districts. Sociopolitical, economic and environmental
aspects will be considered, especially the impact on local smallholders and the
strategies of resistance to oil palm planting.

South Kalimantan

The small province of South Kalimantan continues to support a dense agricultural population, especially in the Hulu Sungai (Figure 6.3). Rice occupies
more than twice as much land as oil palm, though more rice is now grown
in the tidal swamps of Barito Kuala than the Hulu Sungai, and often by
transmigrants. The area under wet rice far exceeds that in the other provinces.

South Kalimantan has traditionally specialized in rubber, but oil palm has been taking over the rubber lands of the northern district of Tabalong, while near Pleihari in the south a former government-owned sugar estate has also been transformed into oil palm, its smallholders being assisted to change crops. A worrying development is the targeting for oil palm of peat swamp areas in the South Hulu Sungai (*Banjarmasin Post*, 2 July 2004, 6 Sept. 2007) and the difficult tidal swamps of Barito Kuala (*Radar Banjar*, 12 June 2006).[18]

Most of the oil palm is grown in the southern district of Tanah Laut or east of the Meratus Mountains, especially in Kota Baru and Tanah Bumbu. The oldest plantation, PT Sinar Kencana Inti Perkasa (now part of the Sinar Mas group), has two mills in operation, while the smallholders, mainly Java-nese transmigrants, control more than 7,000 hectares (*Kapanlagi.com*, 11 July 2007). The south and east coasts of South Kalimantan have numbers of transmigrant settlements, which provide the bulk of the labour force.[19] Local Banjarese are not much involved in such enterprises, but are more likely to work in forestry activities, small-scale farming or illegal coal mining. Meratus Dayak keep well away, living higher up in the mountains, though the protected forest in recent times has been targeted for iron ore and gold mining, prompting the Dayaks to seek assistance from outside NGOs in combating the threat. The "Aliansi Meratus" was formed for this purpose and has campaigned on behalf of the people.

Kota Baru, Tanah Bumbu and Tabalong are coal-mining districts, and it is coal that has created the most serious environmental problems, as well as competing for land with oil palm and local agriculture. Large concessions have been given by the central government to companies such as PT Arutmin (in Kota Baru and Tanah Bumbu) and PT Adaro (in Tabalong). The permit they receive (PKB2B) requires rehabilitation and reforestation of the mine site. Another kind of mining licence (KP) is provided by the district, with no such restrictions. Many miners, known as PETI (*Pertambangan Emas Tanpa Ijin*), are simply illegal. Some work on the actual sites of the large companies, in some cases subcontracting, while others simply dig up the abundant coal resources of the region, which in the Senakin area (Kota Baru) lie close to the surface and often beneath village land. Villagers whose lands were taken over by the large companies received little compensation and only limited amounts of employment, but the illegal miners pay much better (*Kompas*, 17 May 2003, WALHI Kalsel, 30 Aug. 2006). Unfortunately, the results are huge acid-filled holes beyond repair and pollution of watercourses. Illegal coal mining and coal transport have flourished, especially on the east coast, where dozens of small coal ports have sprung up. PT Adaro in Tabalong is Indonesia's largest coal mine. The transport of coal by truck to Banjarmasin has created havoc on the roads (WALHI Kalsel, 14 Apr. 2007).[20] Recently the provincial government decided that oil palm and rubber must become the leading exports, replacing coal, which may soon decline in availability (*Banjarmasin Post*, 15 Jan. 2007).

South Kalimantan may be considered to have two different kinds of "agrarian transition". In the western part of the province, a long-standing "Java-type" transition is in place, with the heavily populated Hulu Sungai continuing to grow rice and rubber but constantly sending (young) people to urban centres such as Banjarmasin, Samarinda and Balikpapan, with the region unable to support its younger generation. The swampy western margins of the North Hulu Sungai, where rice growing is more risky, have become centres for small-scale village industries such as rattan carpet making, batik dyeing and duck raising. Farther south, where the swamps are even deeper and "floating rice" is grown, Negara has an ancient tradition of manufacturing knives and pottery.

From the south and east coasts, through Tanah Laut to Kota Baru and the East Kalimantan border and then continuing north through Tabalong is a frontier region that was heavily forested until the 1980s, with lower population pressure. In this area both mining and oil palm are restricting opportunities for the continuation of traditional agriculture, especially swiddening, rubber growing and collecting of forest products. As more intensive plantation agriculture is established, forests are cleared and the land is occupied by alternative activities. Local villagers are becoming marginalized by these activities, as exemplified by the unfortunate inhabitants of the small island of Pulau Sebuku. Their major problem seems to be the large-scale coal exploitation covering more than half of the island, where a few people have found work as labourers. Most of the inhabitants are fishers and small farmers, and they have seen their environment deteriorate as mangroves (once protected) have been removed and rivers polluted (*Down to Earth* May 2004). They have demonstrated against the slowness of the company in rehabilitating their land, with large holes remaining where they once had rubber groves (*Radar Banjarmasin*, 27 Apr. 2007). In the south, a group of oil palm smallholders on what was once a government property were left stranded by the breakdown of their mill. In 2007 they asked a private iron ore company to take over the land so they could get compensation (*Radar Banjarmasin*, 6 Apr. 2007).[21]

Meanwhile, the push by the provincial government to expand oil palm and rubber at the expense of coal was linked to the "biofuel fever" that gripped the country in 2007. The announcement that 1.1 million hectares would be devoted to biofuel had observers worried about how so much land could actually be found for the product without encroaching on both forest and people's farmland (WALHI Kalsel, 27 Sept. 2007).

Central Kalimantan

Central Kalimantan is Kalimantan's least populated and newest province, having been created only in 1957 to provide a province for Dayaks separate from the Islamic Banjarese of South Kalimantan. The city of Palangkaraya is a similarly recent creation, and quite small (183,000 in 2005), with a less crowded urban environment than Pontianak or Banjarmasin. Palangkaraya's main problem is that it lies inland, away from the major centres of action around Sampit and Kuala Kapuas, and it has very few industries. The timber from the upper reaches of the Barito is handled locally or sent to the sawmills, plywood factories and furniture industries of Banjarmasin.[22] Sampit, one of Indonesia's largest timber ports, deals with production from the rapidly growing centres of Seruyan and Kotawaringin Timur. Many of the oil palm estates are clustered close to the main road between Pangkalanbun and Sampit, at a point where the extensive coastal peats give way to slightly more elevated mineral soil (Figure 6.4).[23]

Previously there were only five districts, plus the city of Palangkaraya: now the five have become 13. Several of the new districts, such as Seruyan, are narrow elongated slivers, their distant northern parts still forest-covered, with small traditional Dayak villages hugging the rivers. Oil palm estates occupy the middle regions, and there are sparsely inhabited swamp forests along the coast, often targeted by illegal loggers. The new districts in particular have been urgently seeking investors, with the result that Central Kalimantan, the site for Suharto's failed million-hectare rice scheme, in May 2004 announced it would establish a million hectares of oil palm. The idea was to place 300,000 hectares in the three districts that now make up Kotawaringin Barat; 400,000 hectares in Kotawaringin Timur, Seruyan and Katingan; and 300,000 hectares in Barito and Kapuas Districts, farther east.

WALHI (Friends of the Earth Indonesia) Kalteng, an umbrella organization representing a number of local NGOs, criticized these plans as threatening forest sustainability. It was expected that the companies would ignore the 4 million hectares of degraded land already available and seek forested land instead. Social problems would be inevitable, as people would be offered inadequate compensation (*Kalteng Pos*, 8 May 2004). The Green Forum was also critical, describing the supposed benefits from the estates as simply myths. Very often locals could find work only as day labourers at meagre wages (*Kalteng Pos*, 14 May 2004). The main reason for the negative reaction towards oil palm in Central Kalimantan has been that the majority of the estates have simply been private enterprises (Perkebunan Besar Swasta, PBS) with no arrangements to include local people. In a detailed list of 56 estates drawn up in 2002, 44 were PBS (Dinas Perkebunan Kalteng 2002). On one estate visited by the author in Kotawaringin Barat, almost the entire workforce consisted of Javanese contract labourers (fieldwork, 1999).

PT Agro Indomas, established in Kotawaringin Timor (now Kabupaten Seruyan) in 1997, engaged in an acrimonious conflict over compensation for 10,000 hectares of land from two villages. The people claimed that much of the land was heavily timbered, providing ironwood for their boatbuilding industry on the shores of Lake Sembuluh. The estate authorities maintained that the vegetation was valueless scrub and grassland. The villagers complained that the company had cleared their rattan gardens, rubber trees and fruit trees, giving them minimal payment, and had occupied the swampland they used for growing rice. In 1999 the villagers destroyed a bridge between the estate and the road. After the district authorities intervened, the company provided compensation but then extended its land claim. An investigation was undertaken by WALHI Kalteng and the British organization Down to Earth. Photographs clearly showed both shipbuilding activity and the remains of forests. The Commonwealth Development Corporation, which had partly funded the company, encouraged it to avoid "insensitive and confrontational" comments and to remain focused on social and environmental issues.[24]

Although the dispute was defused, there were suggestions in May 2006 that the company had not really mended its ways and that agrarian and social conflicts continued between it and the people. Specifically, there were claims that the effluent from the estate factory had polluted Lake Rambania, which could also destroy the fishing industry in Lake Sembuluh (*Kalteng Pos*, 6 May 2006). A year later PT Agro Indomas began construction of a junior high school, free for children of workers on its two estates and available to villagers (*Kalteng Pos*, 23 June 2007). It is not known whether this gesture is related to the fact that the company is now a member of the Roundtable on Sustainable Palm Oil (RSPO), which has laid down strict rules for the way in which estates should relate with surrounding communities.

The Islamic villagers living around the lake suffered severe trauma when the oil palm estates began to arrive in 1994. Their source of livelihood from building ironwood boats was gradually destroyed, together with their traditional subsistence activities of swiddening, fishing and hunting, as their long-standing occupation of the area was ignored.[25] However, with the assistance of WALHI and the NGO they formed in 2003 called Kompak Sembuluh, they have embarked on a range of new activities to demonstrate that they are intensively using the land. They have planted nilam[26] and improved rubber and fruit trees, and are replanting some of the degraded secondary forest. The district authorities were impressed enough to provide a distiller for the nilam, and presumably the villagers are now left in possession of what lands remain under their control (Waseng Vanbroer 2007). They do not work on the estates, which have a transmigrant labour force, but they have independently undergone their own "agrarian transition", as they seek markets for their nilam and eventually increase rubber production.

The reaction in a village in the neighbouring district of East Kotawaringin towards oil palm is a further example of the independent stance now being adopted by some local people, who are exercising agency in their decision making. The village head of Bagendang insisted that his people rejected the advances of an oil palm company because they had already planted cassava to be processed by a nearby factory. It was not possible to overlap with oil palm, and they no longer wanted to be just day labourers on an estate, as it was always people from outside who were preferred. The land they opened for cassava they owned themselves; they planted the crop themselves and sold it directly to the factory without a middleman (*Kalteng Pos*, 26 Aug. 2006).

The district of Seruyan, especially the area around Lake Sembuluh, now ringed with plantations, was at the heart of the ferment over oil palm that gripped Central Kalimantan during 2003 and 2004. The official statistics credited Seruyan with 60,300 hectares in 2003,[27] but the Bupati insisted in February 2004 that at least 250,000 hectares were ready to be harvested while 11 factories had been established and another 6 were planned (*Kalimantan Tengah Dalam Angka* 2003; *Kalteng Pos*, 4 Feb. 2004). A new port was to be constructed for shipping CPO, which, he claimed, would be the largest in ASEAN, or perhaps the world, handling 1.2 million tons per year (*Kalteng Pos*, 4 Feb. 2004; *Kompas*, 18 Mar. 2004). His claims were deflated by the port construction company PT Pelindo III, whose director stated that the company had other priorities for Central Kalimantan (*Radar Banjar*, 9 Dec. 2004).

Another case of the Bupati's misplaced enthusiasm concerned Tanjung Puting National Park, a famous orang utan rehabilitation centre. The park had been invaded by illegal loggers for some years, and much of its timber was degraded (Environmental Investigation Agency-Telapak 2003). The Bupati planned to change the status of 60,000 hectares and invite investors to grow oil palm. He was already at odds with the Forestry Department in Jakarta over the boundaries of three plantations that had strayed into the park, but he was told firmly to cancel any new permits to investors, as the Forestry Department wanted to rehabilitate the area. The Director General of Forest Protection and Nature Conservation reiterated that the Forestry Department was forbidding any further conversions from forest to estate crops (*Media Indonesia*, 22 Feb. 2005; *Kalteng Pos*, 25 Feb. 2005). However, the following year it was revealed that the provincial government had asked for 357,610 hectares of new land in Seruyan to be excised from the forest estate for conversion to oil palm (*Banjarmasin Post*, 28 June 2006). WALHI requested the Minister of Forestry not to allow any more forest conversions, noting that already 35,000 hectares had been converted within the park, an activity that was strictly illegal (WALHI Kalteng, 7 July 2006).[28]

Kotawaringin Timur, the district lying mainly north of the port of Sampit, has also seen rapid oil palm development, with Parenggean sub-district becoming an agribusiness centre. According to 2007 figures, this district had the largest area of planted oil palm in the whole of Kalimantan (199,000 hectares), surpassing the historical leader, Sanggau (West Kalimantan). However, many of the plantings were still immature.[29] The new clearings of forest for oil palm have disturbed numbers of orang utans. The traumatized animals, suffering from loss of habitat, have encroached on the new plantings, eating the pith from young branches. The adults are often killed by the companies, who attempt to sell the young. The local branch of the Balikpapan Orangutan Survival Group has been active in rescuing many animals (*Kompas*, 14 July 2003), and about 200 have now been relocated in a park in the far northern district of Muring Raya (*Banjarmasin Post*, 5 Feb. 2007).[30]

Perhaps because development has been occurring so quickly in Kotawaringin Timur, there is considerable confusion regarding the boundaries of various jurisdictions, with overlaps between mining, village and plantation land, with areas still classified as forest, and between adjacent plantations. As such confusion deters investors, an audit of plantation and other developments was recommended (*Banjarmasin Post*, 29 Apr. 2005, 1 Aug. 2006). One improvement has been that the new estates are more likely to include a smallholder component and to look for a local labour force, as transmigration has declined. While Kotawaringin Timur is second only to Kapuas in population size, such a development is to be welcomed, as it should offer locals more involvement with oil palm than just labouring for a low wage (*Kalteng Pos*, 8 Jan. 2007; *Radar Sampit*, 12 Feb. 2007).

One negative effect of this rapid expansion of oil palm in both Seruyan and Kotawaringin Timur is that much of the new planting encroaches on peat soil. A report by Greenpeace (2008) demonstrates these encroachments through a series of maps, most dramatically around Lake Sembuluh (Seruyan), where one of the major offenders is the Wilmar group, the world's largest producer of palm oil-based biodiesel. The Greenpeace study also criticized the ongoing destruction of orang utan habitat in Central Kalimantan and the continuing practice of estate clearing by burning, even though this is illegal (Greenpeace 2008).

The PLG (Pengembangun Lahan Gambut, or million-hectare peat project) occupies sections of Kapuas, Pulang Pisau and Barito Selatan Districts.

Begun in 1995, it managed to convert only 70,000 hectares of the total 1.4 million hectares into some kind of agricultural activity, mainly based on rice, but in the process it caused massive environmental damage, especially through illegal logging of the swamp forest and burning the peat (*Jakarta Post*, 19 Feb. 2007). About 8,500 transmigrant families remained in the area out of an original 15,000, leading a very marginal existence. They had asked that the area be converted to oil palm, and after several studies it appeared that it would be possible to open 114,000 hectares for that commodity (*Banjarmasin Post*, 26 Feb. 2007). Twelve companies began the process of obtaining permits, but then the central government announced that 80 per cent of the project area (1.1 million hectares) would focus on conservation and restoration of the peat ecosystem, and just 20 per cent on agriculture. The agricultural land would be mainly under wet rice, with only 10,000 hectares available for oil palm, so the oil palm companies would have their permits cancelled (*Banjarmasin Post*, 25 Mar. 2007). These decisions angered the provincial secretary, who predicted that they would impact on the general investment climate of Central Kalimantan. Such comments ignored the difficulties of the deep peat environment. The coordinator of Wetlands International Kalimantan argued that peat was not suitable for growing oil palm: it would be very expensive to make the soil fertile enough for the crop (*Banjarmasin Post*, 25 Mar. 2007). The problems of acid sulphate soils and peat burning had already been raised in Seruyan, where the race to plant more oil palm for biodiesel had brought the peatlands under attention. The level of carbon emissions from the peat made it important that it not be used for oil palm (Nurul Hidayah, *Kalteng Pos*, 17 Feb. 2007). Such conclusions echoed the broader findings of the Dutch study "Peat CO2", which indicated that the reason for Indonesia's third place in global emissions ranking was the result of emissions from drained and burned peatland, which needed to be conserved, not planted (Hooijer *et al.* 2006).[31]

Burning of forest for clearing, and especially peat burning, had made life difficult for the inhabitants of Palangkaraya, which each year seemed to be covered by smoke. Though it was still claimed that swiddeners caused the burning (especially by the central Forestry Department and State Ministry for the Environment), locals knew that certain oil palm estates were largely to blame (*Jakarta Post*, 29 Aug. 2006). These were named in the media and known to the police, but despite threats to cancel permits and fine those found burning, nothing seemed to be done, as the culprits were prominent individuals. Meanwhile, interior people who needed to burn to make swiddens were becoming scared to do so, as in theory they would incur large fines (*Kalteng Pos*, 10 Aug. 2007).

West Kalimantan

Historically Kalimantan's leader in oil palm area and production, West Kalimantan has now lost this prominence. The momentum has slowed, despite government interest in new plantings in the Malaysian borderlands and the push for expanding biodiesel, which began in 2007. Production from the government plantations, already 25 years old, has declined, necessitating large-scale replanting.

The densely populated middle Kapuas basin around Sanggau has the largest oil palm area (around 145,000 hectares in 2007; Figure 6.5), with a mix of government plantations, newer private estates and some independent grower cooperatives. Because of its maturity as an oil palm centre, Sanggau District has also attracted private farms to supply certified seed stock. Its road network, while requiring maintenance, is largely adequate. The road to Sarawak is a vital connection, while the east-west road along the Kapuas River is part of the central link of the Trans-Kalimantan Highway.

Much of the "forest" in this area is the result of human planting, the Dayak villages often possessing extensive tembawang (traditional mixed fruit and timber) gardens in addition to "jungle" rubber and rice swiddens. When the government plantations, now known collectively as PTPN XIII, originally entered the district, the Parindu plantation adopted the "nucleus estate and smallholder model" (NES), whereby the estate supplied 2 hectares of prepared oil palm and a half-hectare house lot to farmers in return for 5 hectares of farmers' land, the balance of which constituted the estate nucleus, or *inti*. Most villagers retained enough land to continue working their rice fields and forest gardens as well as their 2 hectare plots, but as oil palm yields began a premature decline, people began to utilize the estate's fertilizer on their rice fields. Rather than embrace the intensive, high-input "agrarian transition" offered by the estate, they extended their rather low-intensity form of mixed cultivation to incorporate oil palm as one more element (Potter and Lee 1998). Other government estates (such as that at Meliau) were developed without smallholders. They have recently been forced to extend landownership to villagers demanding redress for gardens destroyed and lands resumed. On the government properties, the replanting programme reopened old wounds among villagers, who saw PTPN XIII replanting core estate land without provision for smallholders. Some villagers wanted the land restored to its original owners, while others believed it should be converted into *plasma*, or land occupied by smallholders (*Pontianak Post*, 20 Apr. 2004). Aging trees and declining yields have led to the neglect of oil palm smallholdings in districts such as Landak (*Pontianak Post*, 19 Nov. 2004; *Kalimantan Review*, 12 June 2007). The new government rehabilitation programmes are designed to provide assistance with replanting.

Private companies arriving in Parindu sub-district towards the end of the 1990s had to negotiate directly with individual sub-villages, which they usually did through a high-ranking local official. The success of local people in obtaining a reasonable deal with the plantation depended to a large extent on the quality and attitude of their leaders, with bonuses from the estate authorities to those leaders who could persuade all households to provide the maximum amount of land. One system commonly practised was KKPA (Prime Co-operative Credit for Members). Villagers were asked to supply 7.5 hectares of their land in order to secure 2 hectares of planted oil palm. Bank credit was managed through a cooperative and had to be repaid during the first few years of operation. Our fieldwork in the area in 2002 revealed that those who elected to join a plantation often struggled to find the necessary 7.5 hectares, as they wanted to keep their rice land and tembawang gardens (Potter and Badcock 2007). The tembawang are generally communally owned, and strong adat chiefs often refused to allow them to be cleared.[32] In addition to their age and cultural significance (as sites of former longhouses), they provided an important local source of timber, fruit (especially durians), honey and vegetables. In the case of the Malaysian-owned plantation PT Sime Indo Agro (PT SIA), which we studied, the people had originally rejected the 7.5 hectares/2 hectares arrangement but were forced to comply with the "model" (see also Colchester *et al.* 2006).

Villages that participated in government-sponsored rubber schemes had a further alternative to oil palm, though in 2001, with rubber prices low, the prospect of daily paid estate work was attractive (Potter and Badcock 2007). Dayak rice cultivation, which retains strong cultural significance, is based mainly on wet or dry swiddens. Not only was there pressure on fallow land, but oil palm encouraged rats, which in the early years attacked rice crops as well as palm fruit. During a visit to the area in 2007 the author found that the rat problem had eased. Despite the reduction in their available land, the people had not given up rice growing, though they acknowledged that they harvested only a fraction of what they had once produced. One village had added clonal rubber to its cultivation mix, which had been bought with the proceeds of oil palm. As both rubber and oil palm prices were then high, villagers had increased their incomes and almost paid off their credit. A former agricultural extension officer had bought land in one sub-village; this entrepreneur provided 15 hectares to the company and still had space for cloned rubber (selling seedlings to the villages), fishponds and other initiatives.[33] He believed that PT SIA's presence had led to intensification in farming practices, being more of a catalyst for change in the region than any government agency (Potter and Badcock 2007). Though we found evidence in 2002 of the forms of covert resistance discussed by Scott (1985), such as petty thieving and foot-dragging, people seemed more content with oil palm by 2007 (the high prices surely helping the transition), though some noted the increasing prevalence of gambling and excess alcohol consumption, incipient signs of social decline. Disappointment was expressed at promises made but not kept by the plantation establishment, especially regarding the upkeep of the main road. The breaking of such promises to supply everything from schools to medical facilities was a major source of grievance against the estates (*Kalimantan Review*, 25 May 2007).

A new section of the plantation, recently opened by PT SIA in Sungai Mawang, close to Sanggau, is again causing problems to people who are losing land, including tembawang and rubber gardens (*Kalimantan Review*, 14 Sept. 2007).

The extensive southern district of Ketapang is West Kalimantan's second-largest oil palm region. The plantations are located mainly inland from the swampy coast, in a band running southwards from Gunung Palung National Park . Many transmigrants work as *plasma* (smallholders attached to estates), while Dayak rice crops have been affected by swarms of grasshoppers, which breed in the ubiquitous *Imperata* grasslands. In 1993, when a company began to procure village land in Manis Mata sub-district, near the Central Kalimantan border, its location was certainly remote. The destruction of traditional Dayak livelihoods and the Dayaks' attempts to obtain compensation have been a long-running saga (WALHI Kalbar and Down to Earth 2000).[34] In 2004 some people were still trying to access land under the plantation's contested KKPA scheme and complaining that oil palm had brought disaster to their lives (*Pontianak Post*, 15 Apr. 2004). The *sawitnisasi* (literally "oilpalmization") of Ketapang was regretted by one observer (*Pontianak Post*, 8 Mar. 2005), who believed the costs were much higher than the benefits for ordinary people. Another observer suggested that problems with managing credit, and the buying and selling of allotments encumbered by debt, had become serious. A slowing of the rate of expansion was recommended and a suggestion that independent growers were preferable, as they needed less credit (*Kaltim Post*, 9 June 2004).

West Kalimantan is the founding province of AMAN, the National Alliance of Traditional People (Aliansi Masyarakat Adat Nusantara), originally formed as AMA Kalbar to support local farmers in a land dispute with an oil palm company. It is not surprising that there is debate in Pontianak between those advocating more oil palm (often members of the local parliament, DPRD) and those opposed to its extension. Resistance to the relentless spread of oil palm plantations has been strong among many local people who have struggled to retain aspects of their traditional lifestyle. The levels and types of resistance have grown and changed—partly reflecting the newer political freedoms post-Suharto, but also keeping pace with the changes in the countryside as the industry matures in some areas and the trees become overaged, while in more remote districts the plantations are still new. AMA Kalbar and NGOs such as the Institut Dayakalogy, which publishes the *Kalimantan Review*, recommend alternative tree crops, such as rubber or cocoa. Others are worried about the environmental impact of oil palm on water supplies and the effects of its excessive use of fertilizer and weedicides. One correspondent asked, "Why does it have to be oil palm?" arguing that West Kalimantan did not want to resemble North Sumatra, Riau or Peninsular Malaysia (*Pontianak Post*, 4 Mar. 2005).

One suggestion was "family oil palm" (Kelapa Sawit Keluarga, KSK), which had the support of the DPRD and PTPN XIII. They envisaged a cooperative to prepare village land for oil palm and raise incomes in a project "through the people, by the people, for the people". The palm fruit would be received by PTPN XIII from the districts of Bengkayang, Pontianak and Landak (*Pontianak Post*, 4 Feb. 2005). The plan was to develop 14,127 hectares belonging to 8,972 households. Each household would receive about 1.5 hectares—not enough to live on, but perhaps helpful in improving incomes under mixed cultivation. The representatives of AMA Kalbar, however, remained resolutely opposed, arguing that KSK was a ploy by the company to obtain more land under oil palm without additional costs, all of which had to be borne by the smallholder participants (AMA Kalbar 2005, quoted in *Down to Earth* 66, 2005). The AMA representative stated strongly that oil palm had destroyed the Dayak culture and economy and impoverished the people; she placed special emphasis on the spiritual and ritual significance of the ladang or swidden (*Pontianak Post*, 21 Feb. 2005).

A central government announcement that from 2006 oil palm would be planted along the Kalimantan-Malaysia border brought a reaction from the head of the provincial oil palm planters association. He asserted that the province had been slow in expanding the crop and investors had shown little interest, partly because securing land was perceived as "complicated", the result of lobbying by NGOs, academics and students (*Pontianak Post*, 10 May 2005). The activities of NGOs are still considered largely to blame for the slowness of investment in West Kalimantan, as they encourage communities to reject oil palm (*Harian Equator*, 17 July 2007; *Pontianak Post*, 24 Apr. 2007).

Chinese investors appeared to be interested in the borderlands and were being courted at a time when, according to the chairman of the Chamber of Commerce and Industry, Indonesia desperately needed to expand its stagnant agricultural industry, including the oil palm sector (*Jakarta Post*, 9 June 2005). The central government had been concerned for some time about controlling the Malaysia-Kalimantan border, partly to reduce the rampant illegal logging of this "no man's land" (*tak bertuan*) and in an effort to persuade the population to look to Indonesia rather than Malaysia for employment (*Kompas*, 4 May 2005).

Opposition to the scheme quickly surfaced from many quarters, including the World Wide Fund for Nature (WWF) and the Center for International Forestry Research (CIFOR). Both organizations were concerned over its impact on the international Heart of Borneo conservation initiative. It was argued that it would be better for government to revitalize the existing oil palm area in West Kalimantan: out of 2.3 million hectares planted, 1.5 million had been abandoned (*Pontianak Post*, 16 Sept. 2005).

One plan, drawn up by a group of state-owned plantations, was rejected by the government, especially the Ministry of Forestry, as it envisaged oil palm occupying national parks such as Betung Kerihun and Danau Sentarum (Wakker 2006). Although seven oil palm holdings around Danau Sentarum National Park earlier had their permits cancelled under suspicion of collecting timber, it was feared that virtually all forested land of low to medium elevation would eventually be under consideration from oil palm companies.[35] WWF's international coordinator argued that apart from Danau Sentarum, most of the border area was 1,000 to 2,000 metres high, far above the optimal elevation for oil palm of 200 metres (*Jakarta Post*, 15 Oct. 2005).

The Heart of Borneo initiative was subsequently ratified by Malaysia, Indonesia and Brunei, so these particular national parks are presumably safe from oil palm. By May 2006 the government had retreated from its original proposal, stating that only 180,000 hectares, rather than 1.8 million, had been found suitable for oil palm in the borderlands. It was looking for other available land but making it clear to the districts involved that they should not convert the border forests. The director of the NGO Sawit Watch urged the government to create better regulations for the industry that would benefit local people (*Jakarta Post*, 8 May 2006). Meanwhile, rubber growing has been the dominant industry in the border area of West Kalimantan (*Pontianak Post*, 10 Jan. 2007), though it has been hit badly by declining prices since October 2009.

East Kalimantan

The huge province of East Kalimantan, famous for its oil and gas, forest industries and coal, has made a smaller contribution to oil palm area and production, though not for want of trying (Figure 6.6). Although more forest concessions (Hak Pengusahaan Hutan, or HPH) remain here than in other

Kalimantan provinces, legal forest production has declined. District authorities issued small-scale permits to local cooperatives from 1999 to 2002, resulting in increased clear felling, as villagers seized the opportunity to finally profit from the forests. Illegal logging has now declined, though previously satisfying a demand from local and Javanese industries and for smuggled logs to Sabah. Since the 1980s a large area of logged over, regenerating and degraded forest has existed in East Kalimantan, augmented by the widespread fires of 1998. Potential oil palm companies have followed the prevailing trend and sought good forest as well as transportation access. Relatively high prices for oil palm have produced a boom mentality similar to that of Central Kalimantan, with the usual group of fly-by-night operators.

East Kalimantan's original million-hectare programme, conceived in 1998 by Governor Suwarna, was to construct a kind of "oil palm safety belt" in the border area with Sabah (now Nunukan District). In 2005 the Yudhoyono government was using similar language for its "border protection through oil palm" plan. The earlier scheme was to be developed partly on the extensive Yamaker forest concession, once run by the army, then devolved to the government company Perhutani, but virtually abandoned. There were many young men engaged in illegal logging in this remote location, most of them deported from Sabah, where they already had experience with oil palm (*Suara Pembaruan*, 18 Apr. 2000). The next idea was to spread the notion of "a million hectares" more widely across the province, but to continue to assist returned migrants. In 2002 there was a general call for returned migrants to be given 3 hectares of oil palm per family. It was estimated the province sheltered 300,000 ex-illegal immigrants (*Media Indonesia*, 9 Oct. 2002).

However, the provincial government had already cancelled the location permits of 146 companies, a total of 2.5 million hectares. The story was the common one of companies clearing the forest and taking the timber but not planting. The head of the provincial estate crops office suggested that only 8 per cent of the millions of hectares covered by location permits were serious. He estimated East Kalimantan's million hectares of oil palm would take 15 years to materialize (*Kompas*, 22 Oct. 2003). A year later it was reported that Malaysian investors wanted to plant 1.5 million hectares in the border districts. A Memorandum of Understanding had been signed with Malinau District for 100,000 hectares, Bulungan for a further 100,000 hectares, and Nunukan for 40,000 hectares. However, a new ruling from the National Land Authority allowed foreign private companies a maximum of 20,000 hectares per province (*Kaltim Post*, 7 Oct. 2004).

In February 2005 it was admitted that the million-hectare project had failed, being used only as a way to "spy out" timber. The estate crops office had given location permits to hundreds of large private companies, covering an area of 3.145 million hectares. Only 338,205 hectares were sown, and they became 160,000 hectares of oil palm (*Kompas*, 22 Feb. 2005). East Kalimantan continues to be targeted by oil palm companies, though often for the wrong reasons.

Another problem hindering the rapid spread of oil palm has been the lack of reliable seed. Good planting materials are expensive, Rp13,000 to Rp15,000 per seedling, while the price of poor seedlings is just Rp3,000 to Rp5,000 (*Kompas*, 22 Feb. 2005). Small farmers generally use cheap seeds, which may reduce production levels by 50 per cent and cut the life expectancy of trees. Certified seeds have to come from North Sumatra, which adds to the cost (*Kaltim Post*, 26 Feb. 2005). Such seed problems have been symptomatic of a general shortage: Indonesia must import 20 million to 30 million seeds annually, some from as far away as Costa Rica, until local production is able to meet demand (*Jakarta Post*, 13 Oct. 2005; *Indonesian Commercial Newsletter*, 1 Apr. 2009).

The southern district of Pasir was the first to establish oil palm, and until recently it led the province with the most extensive planted area. Pasir became infamous for a dispute between the company (PTPN XIII) and the indigenous Pasir people, which dragged on for two years. The area in contention was 8,000 hectares of core or *inti* land, which was claimed by 3,000 families from eight villages, who insisted they should acquire the land as *plasma*. The people complained that they had been marginalized: the estate had not improved their lives as promised, and had felled their rattan gardens and fruit trees without compensation (*Suara Kaltim*, 30 Dec. 1999). In December 1999 the villagers invaded the disputed fields and built barricades blocking access roads. The estate closed its nearby factory, making it impossible for its 1,500 transmigrant smallholders to process their fruit. Jakarta authorities instructed the company to give locals a "partnership" (KKPA) scheme and seeds to plant up the land. After further months of wrangling, the estate partially complied. The people claimed the land was theirs under traditional law, but the estate's lawyer saw them as mere shifting cultivators who envied the transmigrant farmers and wanted the land (*Media Indonesia*, 22 Aug. 2001). This dispute was most serious in the damage it caused to social relations, with transmigrants pitted against local villagers.

An old centre like Pasir now has problems similar to those in West Kalimantan, with trees becoming senescent and no longer yielding well. The government's revitalization scheme provided 2.5 million high-quality seedlings to villagers involved in a pilot project of 670 hectares for 2007, the first district selected in East Kalimantan for such a trial, with a larger project (134,418 hectares) for both oil palm and rubber scheduled to be fully operational by 2010. An additional 2,412 hectares of oil palm were to be replanted in ten sub-districts (*Berita, PTPN Nusantara XIII*, 12 Mar. 2007). The plantation's news centre also proudly proclaimed that PTPN XIII would build a biodiesel plant in Tanah Grogot, the first in East Kalimantan, to run next to its cooking oil plant (*Berita, PTPN Nusantara XIII*, 4 Mar. 2007).

By 2007 the ambitious district of Kutai Timur had the largest area of planted oil palm. It was originally caught up in the excitement surrounding the million-hectare scheme, announcing plans to overtake Pasir and become Eastern Indonesia's leading "agropolitan" centre by 2010 (*Kaltim Post*, 28 June 2004). However, a more sober assessment at the time indicated that only ten of the 52 companies granted location permits wanted to continue. Suspicion fell particularly on Malaysian companies that were believed to be only looking for timber (*Kaltim Post*, 23 Feb. 2005). As there were 5,000 unemployed in the district, the Bupati first insisted that each plantation must hire local workers (*Kaltim Post*, 22 Apr. 2004). However, that plan did not last long, and soon Kutai Timur was negotiating to bring in new transmigrants as the investment climate was rapidly changing. By August 2007, the Bupati spoke of 40 companies on a waiting list to obtain permits, with transmigrants coming from Central Java and the victims of the mud volcano in Lapindo (East Java) also invited. The Malaysian interests he had previously rejected (such as Golden Hope) were investing in CPO factories and biodiesel plants, so that the Kombeng-Muara Wahau area (Figure 6.6) was fast becoming the centre of the "new economy" in East Kutai (*Bisnis*, 3 May 2007; *Kaltim Post*, 6 Aug. 2007, 13 Aug. 2007, 15 Aug. 2007). It would appear that the Bupati's ambition for East Kutai has now been fulfilled.

The position in the northern districts of Nunukan and Malinau was rather different. Although Nunukan had some oil palm (and had placed some of the illegal workers deported from Sabah on at least one plantation), it was alleged that companies established near the border with Sabah were engaged in illegal logging and not serious about cultivation, so were very slow to build processing plants (*Radar Tarakan*, 15 Apr. 2004; *Kompas*, 3 Aug. 2004). Part of lowland Nunukan had been scheduled to be incorporated in the Heart of Borneo initiative as a new Sebuku-Sembakung National Park, as it included the habitat of Borneo's endemic pygmy elephant. However, the Bupati was opposed to the creation of this park. The status of the forest was changed and oil palm companies welcomed to Sebuku, only to see their cultivation attacked by elephants (*Radar Tarakan*, 24 Dec. 2005; *Kaltim Post*, 21 Nov. 2006, Sucoco 2009).[36] The extension of the Trans-Kalimantan Highway has improved accessibility to this remote district, but oil palm cultivation has been slow to extend there—though the district authorities have been emphasizing its potential (*Kaltim Post*, 18 Dec. 2009).

Kayan Mentarang National Park, largely located in Malinau District, is an essential inclusion in the Heart of Borneo initiative. While the park is probably protected from oil palm cultivation by its altitude and inaccessibility, Malinau town lies in a valley, and there have been several attempts to establish oil palm on the slopes of some Sesayap River tributaries, including the Malinau River itself. The link to the coast through the Trans-Kalimantan Highway has improved communications here also, although 80 per cent of settlements are still reached only by river or, in parts of Kayan Mentarang, by air (*Kompas*, 17 June 2003). CIFOR scientists have been trying to educate the people and district government of Malinau about the realities of oil palm cultivation and processing. A study tour to Pasir made selected villagers aware of the social problems accompanying the crop (CIFOR 2000). When the district sought an agreement with Sabah Forest Industries to open land for oil palm (*Kaltim Post*, 20 Nov. 2004), CIFOR scientists assessed the resources of a typical forested area likely to be targeted, with steep slopes and thin, easily eroded soil lacking in nutrients (Basuki and Sheil 2005). They advised the Malinau government that oil palm plantations were not sustainable under such conditions, and to their gratification the project was shelved (*Jakarta Post*, 30 Mar. 2005). However, Malinau has the highest poverty levels in East Kalimantan, so in an effort to reduce those, the status of some forest has been changed to allow more land to be used for cropping. It seems that oil palm will finally be established, with possibly three companies being admitted. Though aware of the value of their forest, people are keen for development to take place (*Radar Tarakan*, 23 Jan. 2009, 4 Feb. 2009, 14 Mar. 2009).

The districts that largely constitute the Heart of Borneo area—Malinau (East Kalimantan), Kapuas Hulu (West Kalimantan), Gunung Mas and Murung Raya (Central Kalimantan)—are financially vulnerable. They call themselves "conservation districts", as parks and protected areas occupy much of their land and they nurse the headwaters of major streams. As they are not permitted to cut most of their forests, they are not recipients of the Reforestation Fund (Dana Reboisasi), which supplements the income of other forested districts (*Kompas*, 15 Dec. 2004). Obviously they will need compensation if they are to resist the blandishments of oil palm companies and illegal loggers, but unfortunately, as noted wryly by Moeliono (2006: 7), "The budgeting structure in Indonesia does not reward conservation." The people in all these remote districts still live partly from their swiddens, rubber and forests, though those closer to the border have experience in working periodically in Sarawak, where wages are higher than in Indonesia. They are considered to be poor and backward by planners in provincial centres, but they have managed to keep their cultures relatively intact.

"Sustainable Palm Oil?" New Attempts at Regulation

The Roundtable on Sustainable Palm Oil is a voluntary organization founded in 2003, with support from WWF and mainly European business interests. It is aimed at large companies primarily in Malaysia and Indonesia, together with traders, processors, distributors and financiers. European consumers prefer products certified as coming from sustainable sources, but the industry's environmental and social record has been strongly criticized by NGOs. Two meetings, in Kuala Lumpur and Jakarta, enabled the participants to draw up guidelines for the operation of large estates. The revised version of these was presented at a third meeting in Singapore in November 2005 after extensive public comment (RSPO 2005). The guidelines prohibit plantations from clearing forests, especially High Conservation Value forests; burning is not permitted, and estates are expected to retain or restore biodiversity on and around the property. They must control pesticide use and factory effluents, minimize soil degradation, and maintain quantity and quality of surface and ground water. There must be an assessment of social impacts on local communities and proper systems for dealing with grievances and paying compensation, while employees must receive acceptable pay and conditions.

While it did not seem possible initially to bring smallholders under these rules, in June 2007 a special set of draft guidelines was drawn up to be applied to smallholders seeking certification of their holdings and produce. In the case of smallholders tied to estates, mills and plantations were given three years to bring their smallholders up to the same standards as the core estates: many details of the guidelines were, in fact, directed at estate management. Zen *et al.* (2005) hypothesize that differences in production between smallholders and plantations may be largely due to poorer planting materials and minimal fertilizer use by smallholders. The government's new "revitalization" schemes for estate crops (funded by local banks) are directed especially at oil palm smallholders or *plasma* and, if successful, should assist with the supply of credit for improved seeds and fertilizer. The first two principles of the RSPO guidelines, if adhered to, would remove many of the current difficulties surrounding relationships between smallholders and the plantation. They are (1) commitment to transparency and (2) compliance with applicable laws and regulations (specifically dealing with the right to use land that is "not legitimately contested by local communities with demonstrable rights"). Active participation by smallholders in present and future planning is emphasized throughout the guidelines (RSPO 2007).

Organizations such as Down to Earth have dismissed sustainable palm oil as "mission impossible" and the equating of sustainability with good management as "greenwash" (*Down to Earth* 2004: 1). However, if pressure can be put on companies to change their behaviour or risk losing markets, then there may be positive outcomes. A number of Indonesian estates have now signed up, including some important properties. PT Agro Indomas is a member. Sime Darby has joined, and hence PT SIA and its Indonesian subsidiaries, while in Ketapang the Commonwealth Development Corporation, which owned PT HSL, has sold its holding to agro-industry transnational Cargill, also an RSPO member. Membership does not guarantee compliance with the guidelines, however. One set of estates belonging to the Wilmar group operating in Sambas, West Kalimantan, was recently reported to the RSPO for blatant flouting of guidelines relating to burning as well as treatment of surrounding villagers (Milieudefensie *et al.* 2007).

Conclusion

All indications point to an inexorable increase in areas under oil palm in Kalimantan, as in the rest of Borneo, as this particular type of agrarian transition continues to be supported by both investors and all levels of government. However, the rates of increase in areas under the crop were probably slower in 2009 than in the previous two years as prices were generally lower. The impacts of the demand for biofuel were just beginning to be felt in 2007, but the recent decline in commodity prices may pose questions about the sustainability of using palm oil for biofuel production. At the same time, new plans to revitalize forestry through the REDD initiative may see a move away from clearing mature forest. There is plenty of degraded forest, scrub and grassland theoretically available in Kalimantan, provided that existing claims on the land can be satisfactorily resolved. Indonesia still has a strong advantage for plantation development, with its low labour costs and land availability; and Kalimantan, despite its inadequate infrastructure, is well placed to benefit.

Much of the land included in the Heart of Borneo initiative is theoretically unsuited to oil palm, for reasons of inaccessibility, elevation and poor soils, while the deep peat areas are slowly being recognized as being largely responsible for Indonesia's high emission levels. The "biofuel fever" of 2007 had the effect of pushing up the prices of cooking oil, as companies attempted to export too much of the commodity, without regard for local consumers (*Down to Earth* Aug. 2007). Sensible land-use planning, which would include compensation for districts engaged largely in conservation, should take the pressure off development of oil palm in environmentally marginal areas.

In the remainder of the island, oil palm is a useful product in the overall economic mix, but only if it is indeed produced sustainably, with due consideration to environmental and social factors. A larger role for smallholders in a mixed farming system, rather than the present concentration on plantation monocultures, would be a more suitable goal for future oil palm production in Borneo.

To return to the agrarian transition models of Rigg and De Koninck, it is true that oil palm plantations have produced both agricultural intensification and an increasing integration of production into the market in regions previously characterized by fairly low levels of market involvement. What Rigg has termed "professionalism" may apply to company management practices, although elements of these have been criticized by outside observers. The RSPO, in drawing up a series of guidelines for companies, is both attempting "new forms of regulation", as suggested by De Koninck, and seeking to impose more professional standards, which it is hoped will eventually also be adopted by smallholders. De Koninck's higher levels of industrialization and urbanization and greater population mobility, while applicable to some agrarian transitions, are not directly related to the oil palm story, except in the provision of processing plants and the partial reliance on transmigrant labour. Levels of industry and urbanization, particularly high in East Kalimantan, are related not to agricultural production but rather to by-products of the petroleum industry, such as fertilizer and chemical plants. Such industrial developments attract free migrant labour, as does coal mining in both East Kalimantan and South Kalimantan. De Koninck's final criterion, a revaluation of environmental resources, may be taking place in the area of forestry and includes oil palm as one of its areas of focus. However, it is a fact that agrarian transitions, however described, are patchy in Kalimantan, where swidden farming, inadequate infrastructure and intact traditions continue to characterize the more remote districts.

Notes

1. *Suara Pembaruan*, 18 Apr. 2000; *Kompas*, 22 Feb. 2005. Governor Suwarna was sentenced to an 18-month term in prison and stripped of his position for easing licensing conditions for companies that blatantly exploited East Kalimantan's forests and planted no oil palm (*detik.com*, 20 Feb. 2007).
2. In the late 19th century, trade in forest products experienced a Borneo-wide boom, with a huge international demand for gutta percha as an insulator in undersea cables (Potter 1997).
3. The only exception has been Kutai Barat, which had 31,000 hectares in 2007, almost half the provincial total of 68,000 hectares. In comparison, South Kalimantan

had 184,000 hectares, Central Kalimantan 399,000 (especially along the Barito River) and West Kalimantan 538,000, where Sanggau District alone had 99,000 (*Statistik Perkebunan Kalimantan Barat, Kalimantan Selatan, Kalimantan Tengah dan Kalimantan Timor* 2008).

4. A total of 500 automobiles and 10,000 bicycles were imported from 1923 to 1925.

5. Under these schemes, a farmer group had to provide at least 50 hectares for contiguous planting in a mini plantation.

6. Banjarese were criticized in 1917 for engaging in too many other activities to be considered "serious" farmers, unlike the Javanese (Grijzen 1917).

7. The people affected may include "local" farmers, who could be indigenous or migrants of long standing. Typically, oil palm plantations have also used government-assisted transmigrant labour (most often from Java), while other spontaneous migrants may also be involved.

8. This spontaneous migration into East Kalimantan is continuing (BPS 2006). The province has a large net in-migration, surpassed in national terms only by some districts of Java and Riau plus the Riau Islands in Sumatra. South Kalimantan also has a positive net migration, though smaller, whereas the more agrarian provinces of West Kalimantan and Central Kalimantan both have a negative migration balance.

9. In the early years, many estates simply hired contract labour from Java and did not include smallholders. This was especially the case in more remote areas and common in Central Kalimantan.

10. Apart from East Kalimantan, these rates are still below the Indonesian average.

11. He was also following Dutch colonial practice in attempting to convert "wastelands" into capitalized large-scale production for the market.

12. Note also the larger scale in South Kalimantan (Figure 6.3), which is a much smaller province than the other three.

13. It was quite a difficult process collecting the figures on which these maps are based, so it was not possible to update them further.

14. The Tata Guna Hutan Kesepakatan was the "Agreed Forest Land Use Plan" as set out in the Forest Act of 1967. It consisted of a series of maps, one for each province.

15. Despite such pronouncements, districts continued to seek additional release of forests for oil palm. However, further government statements have followed as Indonesia's acceptance of REDD (Reducing Emissions from Deforestation and Forest Degradation) at the Bali climate conference in December 2007 predicted a decline in levels of deforestation. This was promised in October 2008 as Indonesia committed to a policy of "zero net deforestation" by 2020 at the IUCN World Conservation Congress. Clearing forests for oil palm was specifically targeted (http://www.panda.org/news_facts/newsroom/index.cfm?uNewsID=147348).

16. Kalimantan has about 17 per cent of Indonesia's total oil palm area. Production was an even smaller percentage, with only 11.45 per cent coming from Kalimantan as against 85.55 per cent from Sumatra (United States Department of Agriculture 2009).

17. Some farmers received as little as Rp300–500/kg in October 2008 for their fruit bunches, down from Rp1,900–2,000/kg in September. As many still had to repay debts to their estate, as well as loans taken out when prices were high to buy goods

such as motorcycles, they were having a difficult time (*Koran Indonesia*, 25 Oct. 2008; *Kaltim Post*, 27 Oct. 2008).

18. The potential of swamplands in South Kalimantan (both peat swamps and tidal swamps) that could be used for oil palm has been set at 600,000 hectares by the provincial Plantation Crops Department. This followed a statement by the Minister of Plantation Crops that it was permissible to use such lands as long as they were less than 3 metres deep. Swamps in South Kalimantan are shallow, only 1–1.5 metres deep. Already 100,000 hectares have been reserved for oil palm in the South Hulu Sungai (*Koran Jakarta*, 16 Nov. 2009).

19. An ethnic breakdown of the district of Kota Baru (which at that time included Tanah Bumbu) in the 2000 census revealed that only 37 per cent of the population was Banjarese, with substantial numbers of migrants from Java (27 per cent) and Sulawesi (23 per cent) (BPS 2001).

20. The provincial government acted in 2008 to ban the transport of both coal and oil palm fruit on South Kalimantan's public roads, insisting that a special road needed to be constructed for this purpose. The road is not yet ready, which has held back the recovery of the coal industry, which was hit with low prices during the global financial crisis (Isdijoso 2009).

21. The district council of Kota Baru announced in January 2008 that the defunct palm oil mill on Sebuku Island would be moved to the mainland, where it would be used to produce biodiesel from local palm oil (*Kapanlagi.com*, 17 Jan. 2008).

22. These woodworking industries in Banjarmasin have been suffering raw material shortages and have had to release thousands of employees (*Banjarmasin Post*, 21 Sept. 2006, 25 Aug. 2007, 19 Dec. 2008).

23. The road carries large amounts of traffic. It forms part of the Trans-Kalimantan Highway and used to be in poor condition but has recently been improved.

24. See WALHI Kalteng and Down to Earth 2000; Casson 2001; and Acciaoli 2008: 99 for details on this dispute.

25. Acciaoli (2008) notes that the villages were occupied by Malay Banjarese, rather than Dayak people, when the explorer Lumholtz visited the area in 1920. The Banjarese here may well be descendants of jelutong collectors, who swarmed into the swamplands from the Hulu Sungai during a boom in jelutong, a kind of rubber, in 1908 (Potter 1988, Acciaoli 2008).

26. Nilam is a shrub from which an expensive perfumed oil (patchouli oil) may be extracted. The shrub yields within six months of planting.

27. Official figures for 2005 totalled 65,305. In 2007 the area had almost doubled to 120,426 hectares (Dinas Perkebunan Kalimantan Tengah 2008).

28. The story continues, with the NGO Save Our Borneo accusing the Bupati of corruption as 23 estates (389,680 hectares) that were supposed to be excised from forests to be converted to oil palm belonged to him, to his family or to close associates. In most cases the Minister of Forestry did not permit their excision, but there were questions as to whether the forests had already been cleared. Links between the giant group Wilmar and 18 estates in both Seruyan and neighbouring Kotawaringin Timur (288,480 hectares) were also seen as suspicious (Save Our Borneo 28 Dec. 2009).

29. In 2008 the area of oil palm was given as 322,652 hectares, 203,825 of which were immature (*Kotawaringin Timur Dalam Angka* 2009). This would mean a doubling of the area in one year! It is hard to accept this figure, as Central Kalimantan continues to have many conflicting claims about the actual area under oil palm.

30. The Centre for Orangutan Protection (COP) claims that a Malaysian-owned oil palm plantation encouraged illegal loggers to cut timber in a High Conservation Value forest belonging to a local Dayak village. This forest contained orang utans and 11 other types of mammals and was essential to the lifestyle of the traditional villagers, who grew rattan and rubber and did not want to become oil palm labourers. The loggers began their attack on the forest in October 2008, were briefly chased away by police, but returned between July and October 2009. The COP would like the Roundtable for Sustainable Palm Oil (RSPO) to take up the case (COP 14 Oct. 2009).

31. The Central Kalimantan Peatland Project, with various international contributors— including Wetlands International, WWF and CARE International UK—has been engaged with local communities in restoring the peat swamp environment by closing canals, building small dams and planting local swamp forest trees. The water table has been raised, and carbon sequestration has occurred through tree planting, for which villagers have been paid. The project has reduced emissions from decomposing peat and the incidence of fires, while improving local incomes through the planting of rubber, fruit trees and soybeans (http://www.ckpp.org, http://www. careinternational.org.uk). However, in other parts of the area oil palm estates are apparently still functioning, according to one observer, who contended that the compensation demanded by companies for cancelling their permits was too high for the provincial administration to manage, so the plantations were allowed to continue (Misan 2008). WALHI, the umbrella organization for environmental NGOs, has claimed that the plan to revitalize the peat project has failed (*Suara Pembaruan*, 19 Nov. 2009).

32. "Adat" is traditional law, which is still largely respected in the district.

33. That individual, a Malay from Ngabang, by 2007 had expanded his oil palm to 18 hectares; he also had 9 hectares of *gaharu*, a valuable perfumed wood, as well as cocoa and oranges, plus multiple fishponds (fieldwork, 2007).

34. See Potter (2009) for more information about these disputes.

35. By 2008 oil palm plantations had become established around the border of Danau Sentarum. Greenpeace accused the Sinar Mas company of actually entering the park, building a new road and burning forest (Greenpeace 2008). The impact of pesticide from plantations around the park was being felt in the lake, with its important fishing industry, while sedimentation was increasing (*Media Indonesia*, 31 Jan. 2010).

36. The latest comments, to a local radio reporter by a WWF representative, suggested that 80 per cent of the oil palm cultivation had been destroyed by some of Sebuku's 60 elephants. The district administration was described as "lazy" in not dealing with the ongoing problem (Sucoco 2009).

References

Books, Reports and Journal Articles

Acciaoli, G., "Mobilizing against the 'Cruel Oil': Dilemmas of Organizing Resistance against Palm Oil Plantations in Central Kalimantan", in *Reflections on the Heart of Borneo*, ed. G. Persoon and M. Osseweijer. Wageningen: Tropenbos International, 2008, pp. 91–120.

AMA (Aliansi Masyarakat Adat) Kalbar, "Kronologis kasus gugatan Koperasi Kebun Sawit Keluarga (KKSK) terhadap Aliansi Masyarakat Adat Kalimantan Barat" (The Chronology of the Case against Family Oil Palm According to the Traditional Peoples' Alliance in West Kalimantan), 2005.

Basuki, I. and D. Sheil, "Local Perspectives of Forest Landscapes: A Preliminary Evaluation of Land and Soils, and Their Importance in Malinau, East Kalimantan, Indonesia". Bogor: CIFOR, 2005.

Blommendaal, H.N., *De Oliepalmcultuur in Nederlandsch-Indie*. Haarlem: H.D. Tjeenk Willink & Zoon, 1937.

BPS, *Karakteristik Penduduk Hasil Sensus Penduduk 2000, Kalimantan Selatan*. Jakarta: Badan Pusat Statistik, 2001.

_____, *Karakteristik Penduduk Hasil Sensus Penduduk 2000, Kalimantan Tengah*. Jakarta: Badan Pusat Statistik, 2001.

_____, *Karakteristik Penduduk Hasil Sensus Penduduk 2000, Kalimantan Timur*. Jakarta: Badan Pusat Statistik, 2001.

_____, *Penduduk di Kalimantan Barat*. Pontianak: Badan Pusat Statistik, 2002.

_____, *Estimasi Parameter Demografi: Fertilitas, Mortalitas dan Migrasi. Hasil Survei Penduduk Antar Sensus 2005*. Jakarta: Badan Pusat Statistik, 2006.

_____, *Statistik Indonesia 2007, 2008, 2009*. Jakarta: Badan Pusat Statistik, 2008.

Casson, A., *Decentralization of Policies Affecting Forests and Estate Crops in Kotawaringin Timur District, Central Kalimantan*. CIFOR Reports on Decentralization and Forests in Indonesia, Case Study 5. Bogor: Center for International Forestry Research (CIFOR), 2001.

CIFOR, *Dampak perkebunan kelapa sawit: Wakil masyarakat Hulu Sungai Malinau belajar di Kabupaten Pasir, Kalimantan Timur* (The Impact of Oil Palm Cultivation: People's Representatives from the Upper Malinau River Learn in Pasir District, East Kalimantan). Kabar dari tim pendamping pemetaan desa partisipatif Hulu Sungai Malinau, No. 4. Bogor: Oct. 2000.

Colchester, M. et al., *Promised Land. Palm Oil and Land Acquisition in Indonesia: Implications for Local Communities and Indigenous Peoples*. Forest Peoples Programme, Perkumpulan Sawit Watch, HuMA and the World Agroforestry Centre, 2006.

COP (Centre for Orangutan Protection), "Perkebunan kelapa sawit dan penebang liar adalah kembar siam penghancur habitat orang-utan" (Palm Oil Plantations and Illegal Logging Are the Siamese Twins Destroying Orang Utan Habitat), 2009.

De Koninck, R., "Challenges of the Agrarian Transition in Southeast Asia", *Labour, Capital and Society* 37 (2004): 285–8.

Dinas Perkebunan Kalimantan Barat, *Perkebunan Dalam Angka, Kalimantan Barat 2005*. Pontianak: Dinas Perkebunan, 2006.

_____, *Perkebunan Dalam Angka, Kalimantan Barat 2007*. Pontianak: Dinas Perkebunan, 2008.

Dinas Perkebunan Kalimantan Selatan, *Bahan Pertemuan Sinkronisasi dan Koordinasi Data Perkebunan Kalimantan Selatan di Denpasar 5–9 September 2005*. Banjarmasin: Dinas Perkebunan, 2005.

_____, *Statistik Perkebunan*. Banjarmasin: Dinas Perkebunan, 2007.

Dinas Perkebunan Kaltimantan Tengah, "Investasi Perkebunan 2002" (Investment in Estate Crops 2002), http://www.kalteng.go.id/INDO/Kebun_investor.htm, accessed 29 Mar. 2005.

_____, *Statistik Perkebunan Provinsi Kalimantan Tengah tahun 2005–6*. Palangkaraya: Dinas Perkebunan, 2005–6.

_____, *Statistik Perkebunan Provinsi Kalimantan Tengah Tahun 2007*. Palangkaraya: Dinas Perkebunan, 2008.

_____, "Sejarah perkebunan kelapa sawit di Kalimantan Tengah". Palankaraya: Dinas Perkebunan, 2009.

Dinas Perkebunan Kalimantan Timur, *Statistik Perkebunan 2005*. Samarinda: Dinas Perkebunan, 2006.

_____, *Statistik Perkebunan 2007*. Samarinda: Dinas Perkebunan, 2008.

Dove, M.R., "Smallholder Rubber and Swidden Agriculture in Borneo: A Sustainable Adaptation to the Ecology and Economy of the Tropical Forest", *Economic Botany* 47, no. 2 (1993): 136–47.

Down to Earth, "Dayaks Charged in Oil Palm Dispute", *Down to Earth* 42 (1999).

_____, "Protesters Blockade Australian Coal Mine", *Down to Earth* 61 (2004).

_____, "Indonesia and Biofuel Fever", *Down to Earth* 74 (Aug. 2007).

_____, "The Pressure for REDD: Palm Oil Sector no Longer 'the Golden Crop'", *Down to Earth* 79 (2008).

Elson, R.E., *The End of the Peasantry in Southeast Asia: A Social and Economic History of Peasant Livelihood, 1800–1990s*. London: Macmillan; New York: St Martin's Press, 1997.

Environmental Investigation Agency (EIA)-Telapak, "Update on Tanjung Puting National Park: A Report to the CIG Meeting, Jakarta, December 2003", http://www.salvonet.com/eia/campaigns2_reports.shtml, accessed 10 June 2005.

Greenpeace International, "How Unilever Palm Oil Suppliers Are Burning up Borneo". Amsterdam, 2008.

Grijzen, H.J., *Memorie van Overgave, Zuider en Ooster Afdeeling van Borneo*, no. 272, ARA, Den Haag, 1917.

Hidayati, D., "Striving to Reach 'Heaven's Gate': Javanese Adaptations to Swamp and Upland Environments in Kalimantan". PhD thesis, Department of Geography, Australian National University, Canberra, 1994.

Hooijer, A., *et al.*, *PEAT-CO2, Assessment of CO2 Emissions from Drained Peatlands in SE Asia*. Delft Hydraulics report Q3943 (2006).

Isdijoso, W., "Monitoring the Socioeconomic Impact of the 2008–2009 Global Financial Crisis in Indonesia", Monitoring update. Jakarta: SMERU Research Institute, 2009.

Joekes, A.M., *Koloniaal Verslag* (Appendix C of the Proceedings of the States-General). s-Gravenhage: Algemeene Landsdrukkerij, 1894.

Kalimantan Barat Dalam Angka (West Kalimantan in Figures), 2007.

Kalimantan Review, "Fenomena sawit dan kabut asap di Kalimantan Barat", 25 May 2007.

_____, "Derita petani plasma PTPN XIII Ngabang", 12 June 2007.

_____, "PT SIA gusur tanah, kebun karet dan tembawang masyarakat", 14 Sept. 2007.

Kalimantan Selatan Dalam Angka (South Kalimantan in Figures), 2007.

Kalimantan Tengah Dalam Angka (Central Kalimantan in Figures), 2003.

_____ (Central Kalimantan in Figures), 2006–7.

Kalimantan Timur Dalam Angka (East Kalimantan in Figures), 2007.

Kotawaringin Timur Dalam Angka (East Kotawaringin in Figures), 2009.

Loos, H. and D. van Beusechem, *De Bevolkingsrubbercultuur in Nederlandsch-Indie III. Westerafdeeling van Borneo*. Weltevreden: Landsdrukkerij, 1925.

Luytjes, A., *De Bevolkingsrubbercultuur in Nederlandsch-Indie II. Zuider-en Ooster Afdeeling van Borneo*. Weltevreden: Landsdrukkerij, 1925.

Mayer, J., "Long Ikis: Rattan vs Oil Palm", Typescript report for Institute of Current World Affairs, Hanover, 1988.

Milieudefensie, Lembaga Gunawan and Kontak Rakyat Borneo, *Policy, Practice, Pride and Prejudice: Review of Legal, Environmental and Social Practices of Oil Palm Plantation Companies of the Wilmar Group in Sambas District, West Kalimantan (Indonesia)*. Amsterdam: Milieudefensie (Friends of the Earth), 2007.

Misan, "Ancaman deforestasi dan kerusakan lahan gambut di tengah pembangunan perkebunan kelapa sawit Kalimantan Tengah", www.satuportal.net, accessed 14 Nov. 2008.

Moeliono, M., "Wilderness and Order: Forest Conservation in Malinau District, East Kalimantan". Paper presented at "Survival of the Commons: Mounting Challenges and New Realities", 11th conference of the Association for the Study of Common Property, Bali, Indonesia, 18–23 June 2006.

Oil World Annual. Hamburg: ISTA Mielke GmbH, 2009.

Ozinga, J., *De Economische Ontwikkeling der Westerafdeeling van Borneo en de Bevolkingsrubbercultuur* (The Economic Development of the Western Division of Borneo and Smallholder Rubber Cultivation). Wageningen: Zomer en Keunig, 1940.

Padoch, C., E. Harwell and A. Susanto, "Swidden, Sawah and In-between: Agricultural Transformation in Borneo", *Human Ecology* 26 (1998): 3–20.

Perusahaan Negara Perkebunan (PNP) VII, *Pertama di Kalimantan Barat: Pabrik Kelapa Sawit Gunung Meliau* (First in West Kalimantan: Gunung Meliau Palm Oil Mill), "Proyek Pengembangunan Kelapa Sawit di Kalimantan Barat" (Oil Palm Plantation Development in West Kalimantan), 1984.

Potter, L., "Indigenes and Colonisers: Dutch Forest Policy in South and East Borneo (Kalimantan), 1900 to 1950", in *Changing Tropical Forests: Historical Perspectives on Today's Challenges in Asia, Australasia and Oceania*, ed. J. Dargavel, K. Dixon and N. Semple. Canberra: Centre for Resource and Environmental Studies, Australian National University, 1988, pp. 127–53.

_____, "Environmental and Social Aspects of Timber Exploitation in Kalimantan, 1967–89", in *Indonesia: Resources, Ecology and Environment*, ed. J. Hardjono. Kuala Lumpur: Oxford University Press, 1991, pp. 177–211.

————, "Banjarese in and Beyond the Hulu Sungai: A Study in Cultural Independence, Economic Opportunity and Mobility", in *New Challenges in the Modern Economic History of Indonesia: Proceedings of the First Conference in Indonesia's Modern Economic History*, ed. J.T. Lindblad. Leiden: Bureau of Indonesian Studies, Leiden University, 1993, pp. 264–98 (also in Indonesian).

————, "A Forest Product out of Control: Gutta Percha in Indonesia and the Wider Malay World, 1845–1915", in *Paper Landscapes: Explorations in the Environmental History of Indonesia*, ed. P. Boomgaard, F. Colombijn and D. Henley. *Verhandelingen van het Taal- Land en Volkenkunde* no. 178, Leiden (1997): 281–308.

————, "Oil Palm and Resistance in West Kalimantan, Indonesia", in *Agrarian Angst and Rural Resistance in Southeast Asia*, ed. D. Caouette and S. Turner. London and New York: Routledge ISS Studies in Rural Livelihoods, 2009.

Potter, L. and S. Badcock, "Can Indonesia's Complex Agroforests Survive Globalization and Decentralization? Sanggau District, West Kalimantan", in *Environment, Development and Change in Rural Asia-Pacific: Between Local and Global*, ed. J. Connell and E. Waddell. London and New York: Routledge, 2007.

Potter, L. and J. Lee, *Tree Planting in Indonesia: Trends, Impacts and Directions*, Occasional Paper No. 18. Bogor: Center for International Forestry Research, 1998.

Proyeksi Penduduk 2000–2025: 3.5. Urbanisasi, http://www.datastatistik-indonesia. com/proyeksi/index.php.

Rigg, J., "Poverty and Livelihoods after Full-time Farming: A South-east Asian View", *Asia Pacific Viewpoint* 46, no. 2 (2005): 173–84.

Rigg, J. and S. Nattapoolwat, "Embracing the Global in Thailand: Activism and Pragmatism in an Era of Deagrarianisation", *World Development* 29, no. 6 (2001): 945–60.

RSPO (Roundtable on Sustainable Palm Oil), "Briefing Note: Key Issues Arising from the First Public Consultation of the RSPO Principles and Criteria for Sustainable Palm Oil". Prepared for the second meeting of the RSPO Criteria Working Group, 14–18 Feb. 2005.

————, *RSPO Principles and Criteria for Sustainable Palm Oil Production: Consultation Draft Guidance on Smallholders*. Prepared for comments from the Task Force on Smallholders, 15 June 2007, ed. M. Colchester. Forest Peoples Programme (www. rspo.org).

Save Our Borneo, "Aroma korupsi alih fungsi hutan untuk kelapa sawit di Seruyan", Indonesian Facebook and other websites, accessed 28 Dec. 2009.

Scott, J.C., *Weapons of the Weak: Everyday Forms of Peasant Resistance*. New Haven: Yale University Press, 1985.

————, *Domination and the Arts of Resistance: Hidden Transcripts*. New Haven and London: Yale University Press, 1990.

Statistik Indonesia 2009, www.bps.go.id, accessed 13 Feb. 2010.

Sucoco, "Pemda Nunukan lamban atasi konflik gajah vs manusia", http://www.greenradio. fm/index.php?option=com_content&view=article&id=1581, accessed 20 Nov. 2009.

Thee Kian-Wee, *Plantation Agriculture and Export Growth: An Economic History of East Sumatra, 1863–1942*. Jakarta: LEKNAS-LIPI, 1977.

United States Department of Agriculture, "Indonesia: Palm Oil Production Growth to Continue", 19 Mar. 2009.

Vidyattama, Y., H. Hill and B. Resosudarmo, "Indonesia's Changing Economic Geography". Presented at seminar, Research School of Pacific and Asian Studies, ANU, 22 May 2007.

Wakker, E., *The Kalimantan Border Mega-project*. Friends of the Earth Netherlands and the Swedish Society of Nature Conservation, 2006.

WALHI Kalimantan Barat and Down to Earth, *Manis Mata Dispute*. Jakarta: National Executive WALHI, 2000.

WALHI Kalimantan Selatan, "The Dark Picture of Coal Mining in South Kalimantan (6) Land and Social Conflicts", 30 Aug. 2006.

_____, "Hentikan jalan umum untuk angkutan batubara", 14 Apr. 2007.

_____, "Menyikapi ekspansi perkebunan kelapa sawit untuk biodiesel", 27 Sept. 2007.

WALHI Kalimantan Tengah and Down to Earth, "The Dispute between the Local Community and PT Agro Indomas Oil Palm Plantation, Central Kalimantan, Indonesia", Sept. 2000.

WALHI Kalimantan Tengah, "Environmental Outlook WALHI Kalimantan Tengah 2004: Perilaku biadab terhadap Lingkungan dan Sumber Kehidupan Rakyat", http://www.walhi.or.id/kampanye/globalisasi/utangeko/041203_envoutlook_wkalteng.

WALHI Kalimantan Tengah, "WALHI Kalimantan Tengah desak Menhut dan Guburnur Kalteng untuk tidak memberikan izin pelepasan kawasan hutan bagi perkebunan sawit di Kabupaten Seruyan", 7 July 2006.

Ward, M.W. and R.G. Ward, "An Economic Survey of West Kalimantan", *Bulletin of Indonesian Economic Studies* 10, no. 3 (1974): 26–53.

Waseng Vanbroer, "Menguasai wilayah Kelola dengan Nilam". Report of Arie Rompas, 12 Feb. 2007, http://waseng-vanbroer.blogspot.com, accessed 6 Mar. 2007.

World Bank, *Indonesia: Strategies for Sustained Development of Tree Crops (Rubber, Coconut and Oil Palm)*. Washington, D.C.: World Bank, 1989.

Zen, Z., C. Barlow and R. Gondowarsito, "Oil Palm in Indonesian Socio-economic Improvement: A Review of Options". Paper presented in seminar series, Indonesia Study Group, Australian National University, 2 Mar. 2005.

Newspaper Articles

Banjarmasin Post

8 May 2002   – "Nasib petani di lumbung beras Kalsel"
2 July 2004   – "HSS manfaatkan rawa untuk sawit"
29 Apr. 2005 – "Tumpang tindih hambat investasi"
28 June 2006 – "Pembukaan lahan penyebab banjir"
1 Aug. 2006   – "Audit perusahaan sawit"
21 Sept. 2006 – "Mengulang kejayaan industri perkayuan mungkinkah?"
15 Jan. 2007 – "Perkebunan gantikan batu bara"
5 Feb. 2007   – "Orangutan segara dilepas"
26 Feb. 2007 – "114,100 hektare PLG untuk sawit"

25 Mar. 2007 – "Lahan sawit hanya 10,000 hektare"
25 Aug. 2007 – "Pengadilan hubungan industrial sebuah pilihan yang berat"
6 Sept. 2007 – "Investor hanya garap lahan rawa"
19 Dec. 2008 – "Ribuan petani Kalsel dirumahkan lagi"

Berita, PTP Nusantara XIII

4 Mar. 2007 – "PTPN XIII juga akan bangun pabrik minyak goreng di Paser"
12 Mar. 2007 – "Pemkab Paser kembangkan 2.5 juta bibit sawit"

Bisnis

3 May 2007 – "3 investor bangun pabrik biodiesel"

detik.com

20 Feb. 2007 – "Citra satelit ungkap korupsi"

Harian Equator

17 July 2007 – "Stop izin kebunan sawit"

Indonesian Commercial Newsletter

1 Apr. 2009 – "Plant Seedling Cultivation Industry"

Jakarta Post

30 Mar. 2005 – "Future Rides on Land Use"
9 June 2005 – "Chinese Investors Eye RI Oil Palm Sector"
13 Oct. 2005 – "No Changes in CPO Export Duty, Yet"
15 Oct. 2005 – "WWF Warns Govt over New Plantation Areas"
8 May 2006 – "Govt Seeks New Land for Border Project"
29 Aug. 2006 – "Will Indonesia Ever Stop Exporting Haze?"
19 Feb. 2007 – "Peatland Project to Focus Mainly on Conservation"

Kalteng Pos

4 Feb. 2004 – "Seruyan bangun pelabuhan CPO terbesar di ASEAN"
8 May 2004 – "1 juta ha kebun sawit ancam kelestarian hutan"
14 May 2004 – "Pemerintah terjebak mitos kebaikan perkebunan"
25 Feb. 2005 – "Menhut akan ke Kalteng"
6 May 2006 – "Pencemaran Danau Rambania oleh PT Agro Indomas"
26 Aug. 2006 – "Warga Bagendang Permai tolak sawit"
8 Jan. 2007 – "Dua tahun realisasi SHU Rp 7.7M"
17 Feb. 2007 – "Meminpikan kesejahteraan dari kelapa sawit"
23 June 2007 – "Peduli pendidikan, PT AI bangun SMP"
10 Aug. 2007 – "Perlu solusi soal pembakaran lahan"

Kaltim Post

22 Apr. 2004 – "Kelapa sawit di Kutim jadi primadona"
9 June 2004 – "Jual beli kavling sawit makin ramai"

28 June 2004 – "Kaltim mau jadi pusat agroindustri"
7 Oct. 2004 – "Investor Malaysia minta 1.5 juta ha"
20 Nov. 2004 – "SFI Malaysia sudah tandatangani MOU"
23 Feb. 2005 – "Malaysia pun incar Kutim dudung: Kami prioritaskan pengusaha dalam negeri"
26 Feb. 2005 – "Socfindo cari 20 ribu hectare lahan"
21 Nov. 2006 – "Kedati bidik petinggi Nunukan"
6 Aug. 2007 – "Kutim targetkan 500 ribu hektare"
13 Aug. 2007 – "Rp 8M berputar di Kongbeng"
15 Aug. 2007 – "Siap bangun industrial estate dari hasil CPO"
27 Oct. 2008 – "Kapan harga … harus turun"
18 Dec. 2009 – "TKI cukup garap lahan di Nunukan"

Kapanlagi.com

11 July 2007 – "Petani plasma sawit Kotabaru ketiban rejeki kenaikan harga migor"
17 Jan. 2008 – "Kotabaru produksi biodiesel 6 ton/hari"

Kompas

17 May 2003 – "Tambang batu bara illegal di Kota Baru"
17 June 2003 – "Kabupaten Malinau"
14 July 2003 – "Perkebunan kelapa sawit di Kalteng ancaman serius habitat orang-utan"
16 Sept. 2003 – "Menanti investor sawit yang peduli petani"
22 Oct. 2003 – "Izin lokasi 146 perusahaan perkebunan dicabut"
18 Mar. 2004 – "Kabupaten Seruyan"
3 Aug. 2004 – "Tak ada pabrik pengolahan, petani kesulitan manjual kelapa sawit"
15 Dec. 2004 – "Daerah konservasi mestinya dapat kompensasi"
22 Feb. 2005 – "Proyek sejuta hectare kebun sawit gagal, lahan ditelantarkan"
4 May 2005 – "Perbatasan, sejengkal wilayah tak bertuan"
13 Mar. 2007 – "Investasi sawit di Kalimantan macet"
10 Apr. 2007 – "Konversi lahan dilarang"
6 June 2007 – "Karet rakyat, petani terimpit di Hulu dan Hilir"

Koran Jakarta

16 Nov. 2009 – "Potensi sawit rawa di Kalsel 600,000 hektare"

Koran Indonesia

25 Oct. 2008 – "Masih sulit petani sawit"

Media Indonesia

22 Aug. 2001 – "Polda diminta tindak penutup PTPN XIII"
9 Oct. 2002 – "Depnakertrans minta lahan sawit untuk TKI"
22 Feb. 2005 – "Pemerintah tetap larang konversi hutan ke perkebunan"
25 May 2009 – "Polda Kalsel larang angkutan baru bara lewat jalan negara"
31 Jan. 2010 – "Ekspansi sawit percepat sedimentasi Danau Sentarum"

Pontianak Post

15 Apr. 2004 – "Kami mahasiswa selaku warga Manis Mata"
20 Apr. 2004 – "PTPN XIII ingkari hasil kesepakatan"
19 Nov. 2004 – "Peremajaan kelapa sawit"
4 Feb. 2005 – "Manfaatkan lahan rakyat untuk sawit"
21 Feb. 2005 – "AMA Kalbar tolak program kebun sawit keluarga"
4 Mar. 2005 – "Kelapa sawit adalah musuh tanah dan air"
8 Mar. 2005 – "Sawit adalah bisnis pejabat kapitalis"
10 May 2005 – "Buka lahan sawit sepanjang perbatasan"
16 Sept. 2005 – "Kedok untuk ambil kayu: 1.5 juta ha sawit tidak aktif"
30 Dec. 2006 – "Harapkan daerah perbatasan jadi sentra perkebunan karet"
10 Jan. 2007 – "Karet dominasi Perbatasan"
28 Feb. 2007 – "Karetisasi, Kalbar butuh dana Rp 6.5M"
24 Apr. 2007 – "Stop konsensi baru"

Radar Banjar(masin)

9 Dec. 2004 – "Seruyan bangun CPO terbesar di dunia"
12 June 2006 – "IP dan lokasi PT PBB telah diproses"
6 Apr. 2007 – "Perkebunan sawit belum memberi hasil: Warga mengininkan per-
              kebunan sawit jadi tambang biji besi"
27 Apr. 2007 – "PT BCS didemo warga 4 desa tuntut reklamasi lahan tambang"

Radar Sampit

12 Feb. 2007 – "PBS munculkan ekonomi kerakyatan"

Radar Tarakan

15 Apr. 2004 – "Kadishutbun dipanggil polda jadi saksi, terkait perizinan alat berat
              PTNJL"
24 Dec. 2005 – "Kawasan gajah mengamuk di Sebuku"
23 Jan. 2009 – "Warga tak mau cekewa lagi. Dinas Perkebunan jamin komitmen PT
              JPI"
4 Feb. 2009 – "Pemkab yakin PT JPI angkat kesejahteraan warga"
14 Mar. 2009 – "Lima perusahaan tunggu izin pendaratan alat"

Suara Kaltim

30 Dec. 1999 – "Perkembangan aksi kegiatan pemasangan Portal di perkebunan sawit
              inti yang deklaim masyarakat 8 desa"

Suara Pembaruan

18 Apr. 2000 – "Kaltim akan buka 1 juta hectare perkebunan sawit di perbatasan"
19 Nov. 2009 – "Gagalnya revitalisasi lahan gambut di Kalteng"

Tribun Kaltim

26 Apr. 2009 – "Nilai ekspor plywood anjlok 54 persen"

Borneo in the Eye of the Storm and Beyond

Rodolphe De Koninck

Borneo's Own Agrarian Transition

As elsewhere in the world, and in other parts of Southeast Asia, resource extraction has a long history in Borneo. But since the 1960s, it has gained momentum with timber extraction and, more recently, with the expansion of boom crops, particularly oil palm. These two often-interrelated processes have accompanied and, more often than not, generated and even spearheaded a series of landscape and socio-economic transformations. More important, as demonstrated by the contributors to this volume, none of this has occurred uniformly and unilaterally nor without unexpected resistance as well as adaptations. The more salient features that emerge from the contributors' analyses can be summarized as follows.

(1) Although expansion of boom crops has recently become the leading process, with rubber and particularly oil palm being favoured, other forms of resource extraction remain, while urban development proceeds rapidly and not always in direct relation with agricultural growth.

(2) Agricultural expansion does not necessarily occur in lieu of intensification but rather in combination with it, as the experience of Peninsular Malaysia began to illustrate in the 1960s and 1970s. There, the development by the Federal Land Development Authority (FELDA) of large smallholder settlement schemes, devoted initially to rubber then increasingly to oil palm, was accompanied by significant increases in crop yields. Thus was tapped smallholders' traditional propensity to produce more from the soil than large-scale operators, a feature more characteristic of rice cultivation (Gourou 1940, De Koninck 1983, Bray 1986). With reference to Sarawak,

and as specifically emphasized by Cramb (Cf. Chapter 3), not only can agricultural expansion and intensification be associated, but "waves of expansion and contraction [can occur], within which are found pockets of intensification and 'disintensification'". Cramb also demonstrates the intricate forms of agrarian transition that have characterized Sarawak, where actual phases or stages of such transition have taken place, "resulting in a series of partially overlapping and mutually determining (or 'imbricated') socio-ecological landscapes".

(3) Of course, expansion and intensification have not necessarily proceeded—or receded—in the same manner throughout the big island. Between regions, time frames have been different and so have the commercial crops involved. For example, tobacco in Sabah (North Borneo) and pepper in Sarawak have played a key role during expansion phases, in the case of the former as far back as the 1880s. In Kalimantan, where a number of such crops have been involved depending on the region, it is generally rubber cultivation that, almost to this day in some provinces, has acted as the expansion spearhead. In 2007, as Lesley Potter shows (Cf. Chapter 6), it was still the most widespread cash crop in West Kalimantan. As for oil palm, which has become the dominant crop throughout most of the rest of the island, its expansion began much earlier in Sabah than in Sarawak and Kalimantan. However, by the last decade of the 20th century a convergence of choices had become evident, with its cultivation expanding rapidly in all states and provinces of Borneo, placing the big island at the centre of the Southeast Asian oil palm boom (Table 2.4).

(4) The rapid territorial expansion of commercial crops has had a direct demographic impact, quantitatively, qualitatively and spatially. Official demographic statistics do indicate that overall population growth in Borneo has been moderately above both Malaysian and Indonesian averages. If estimates for the number of illegal or non-registered immigrants are also taken into account, Borneo's population growth appears substantially above average. Although not exclusively and directly related to agricultural expansion and growth, it is largely attributable to them. Nearly everywhere on the island, most of the expansion has occurred on large plantations that rely predominantly on migrant labour or, as Potter shows in the case of Kalimantan, transmigrant labour. However, not all migrant labourers come from outside Borneo Island, as plantations in both Sabah and Sarawak employ migrants coming in from Kalimantan. Whatever the case, Borneo's ethnic structure is transformed at all levels. As has been and continues to be the case in other frontier areas of Southeast Asia, such as Mindanao Island in the Philippines and the Central Highlands of Vietnam, the share of

indigenous populations in the total population is decreasing. Within specific areas, such as the provinces of Kalimantan, specific migrant groups are by now particularly well represented (Table 6.2). Overall, as its population increases, the big island's ethnic composition is becoming more diversified and complex, as Potter clearly shows (Cf. Chapter 6). Concurrently, its distribution is also modified in at least two ways. As agricultural expansion proceeds towards the interior of the island, it contributes to the latter's population growth. And, as resource extraction continues or increases, whether of timber, boom crops or minerals—in the latter case, particularly in East Kalimantan—it contributes to the development of trade and of the urban network. In the meantime, the major cities, still predominantly coastal, consolidate or increase their demographic and economic pre-eminence. Overall, in-migrations, whether or not officially sponsored, favour the plantation sector and thus represent a clear development choice. Consequently, as Cramb rightly points out, agricultural expansion appears not only as an economic process but also as a political one that brings about profound socio-economic transformations (De Koninck 2006).

(5) Until the end of the 20th century, agricultural expansion was largely state-driven, in both Indonesia and Malaysia—in this case largely through FELDA—and the reduction of poverty represented not only an official but also a genuine goal. However, since then, as amply demonstrated in Borneo, boom crop expansion has not been driven primarily by social policy, as it has gradually been taken over by private corporations. The latter's goals have little to do with poverty reduction among local populations. Even if agricultural expansion has predominantly become the business of large-scale plantations, often relying on the help of the state, a large number of smallholders do get involved and reap benefits from cultivating oil palm on their own plots. Cramb (Cf. Chapter 3), Bissonnette (Cf. Chapter 4) and Potter (Cf. Chapter 6) demonstrate the resilience of smallholders who adapt to the new boom crop—as they did in the past to other boom crops—cultivating it on their own but bringing their harvested fruit to the estates' palm oil mills for processing. Nevertheless, many others contest and resist the encroachment of large plantations, which tend to expand their operations through various forms of land grabbing—without, or more frequently with, the official or unofficial support of the state or provincial authorities. As amply shown by Cramb as well as Bissonnette about Sarawak, Bernard and Bissonnette about Sabah, and Potter about South Kalimantan, these authorities go to all lengths to facilitate the development of private and particularly large plantations.

(6) For this to be achieved, these plantations must gain access to land, to much land. That is why, in the first place, they are investing in Borneo,

where forestland is abundant and supposedly available for development. Unsurprisingly, competition for land appears as one of the key issues, in a context where both the state and private corporations frequently challenge the land tenure status of local communities. As Bissonnette, Cramb and Potter each demonstrate, local populations' reactions to the oil palm boom, or at least to the one spearheaded by private plantations, have often taken the form of resistance, at times violent. This seems to be particularly common in Sarawak, where recent changes in land laws by the state government, often verging on outright manipulation, render land claims by the indigenous populations particularly difficult. But that does not prevent them, particularly the Ibans, from resorting to the courts to claim their right of access to native customary land.

(7) With reference to agricultural expansion at the expense of forestland, the territorialization role of the state has historically been much more explicit in Malaysia, particularly through the above-mentioned FELDA agency and, more recently, by the state governments of Sabah and Sarawak. But even on the Indonesian side, as shown by Potter with reference to the province of South Kalimantan, the provincial state is getting directly involved in facilitating the replacement of coal by rubber and oil palm as leading exports, largely through the development of private plantations. This is often achieved through a considerable wastage of forestland. As Potter reveals, after 2.3 million hectares of land were planted with oil palm in West Kalimantan during the early part of the 21st century, 1.5 million hectares were abandoned. In East Kalimantan, "the story was the common one of companies clearing the forest, taking the timber and not planting".

(8) Overall, the peasantry's own territoriality is increasingly questioned by plantation expansion, which, as Bissonnette suggests, is likely to eradicate the direct link between peasant households and territorial-economic management. However, as he also shows, it does not necessarily imply that these households give in and abandon their land. In fact, communities have "opted for an endogenous initiative to achieve economic development based on agriculture, without even requesting legal individual land titles". This amply corroborates what both Cramb and Potter show regarding peasant resilience and initiative.

(9) With reference to the agrarian transition occurring in Borneo, Cramb aptly points out that "expansion of the agricultural frontier is the key process in Sarawak's agrarian transition, converting forestland to modern plantation agriculture on an unprecedented scale". This can also be said about Sabah and Kalimantan, where the agrarian transition—or transitions, as Potter specifies—also essentially fuelled by agricultural expansion, appears "highly

problematic and contested", to use Cramb's own terms. Considering the local provincial and state authorities' grandiose plans for continuous oil palm expansion, Borneo's own agrarian transition will continue to bring about the exclusion but also the resilience and adaptation of its inhabitants, whether indigenous or migrants. And along with these processes, the global integration of the big island at the centre of Southeast Asia is likely to be strengthened and the vulnerability of its environmental and cultural heritages aggravated.

Oil Palm Expansion beyond Borneo?

Notwithstanding the increasingly alarming testimonies concerning the demise of Borneo's forest heritage (Borneo Sustainability Forum 2009), signs are that the massive deforestation and oil palm expansion storm that have been engulfing the big island for several decades are far from over. To this must be added a new form of onslaught on peat swamp forest, where, as in many other regions of Southeast Asia, shrimp aquaculture is fast being developed (Langner *et al.* 2007). But there are also signs that the oil palm storm is expanding beyond Borneo, beyond Southeast Asia itself, to several other regions of the tropical world.

Cultivation of oil palm will remain contentious, but considering market demand and, equally important, nearly universal state support, it is likely to continue to expand, on both smallholdings and, even more, large plantations. This applies both to Sabah and Sarawak's contribution to the Malaysian oil palm sector and to Kalimantan's own contribution to the Indonesian one, in a context where the two countries still supply nearly 90 per cent of world demand for oil palm products. Such a commercial success is largely attributable to the fact that, over recent decades, and contrary to most boom crops, palm oil has shown remarkable resilience to market fluctuations. Although prices did collapse along with those for fossil fuels in 2008, they have since rebounded and palm oil futures appear promising. This is largely attributable to the diversity in demand sources. Palm oil is not only the world's second-most important edible oil after soybean oil, it is increasingly sought after as a component for various food products and cosmetics, soap in particular. In addition, it is now a key source of biodiesel, a commodity that, given the popularity of so-called green fuels, is very likely to gain market share, whatever its own environmental production costs, including the depletion of overall forest resources. In addition, given the high profitability of the crop, funds are abundant for research, which continues to contribute to increases in yield as well as improving the resistance of oil palm trees to various actual and potential diseases.

While some of the big Malaysian oil palm corporations are currently investing in the crop's expansion in Cambodia, the outcome is unlikely to be sufficient to meet the continuing increase in world demand. In fact, these same corporations are also investing massively in Papua New Guinea, tropical Africa and Latin America, notably Colombia, while China is also becoming an important producer. This might, in turn, reduce the pressure on Southeast Asia, particularly on Borneo, currently the region's most intensive oil palm expansion laboratory. In such a case, will the respective Malaysian and Indonesian state and provincial authorities take the opportunity to look closely into the possibility of subscribing to the REDD programme (Reducing Emissions from Deforestation and Forest Degradation in developing countries) (Angelsen *et al.* 2009)? This yet-to-be-applied UN initiative is intended to "create a financial value for the carbon stored in forests, offering incentives for developing countries to reduce emissions from forested lands and invest in low-carbon paths to sustainable development". This appears particularly relevant to Borneo, whose "natural forest ecosystems [...] are one of the carbon-densest forests and store a considerable fraction of terrestrial carbon of the world. They also harbour a rich assemblage of biological diversity and provide a range of services for human well-being both in a local and global context" (Borneo Sustainability Forum 2009).

If the REDD programme were to become reality, it might provide Malaysia and Indonesia with an alternative to the overexploitation of the forest resources of Borneo and the excessive expansion of oil palm, through the establishment of extensive permanent forest estates to be sustainably managed to harmonize human uses with carbon stock and conservation. However, for the moment, given the limited results of the December 2009 Copenhagen Conference on Climate Change to promote agreements for a better international management of the environment, and the so far very cautious statements of the Malaysian and Indonesian authorities on the REDD proposals (Parker *et al.* 2008: 42–3), it is difficult to foresee the outcome.

References

Angelsen, A. *et al.*, "Reducing Emissions from Deforestation and Forest Degradation (REDD)". An Option Assessment Report produced for the Government of Norway, Meridian Institute, http://www.REDD-OAR.org, 2009.

Borneo Sustainability Forum, "Suggestions Derived from the Joint Activities of the Borneo Sustainability Forum", Nov. 2009, http://www.bakosurtanal.go.id/upl_images/Borneo%20mapping%20forum-darmawan.pdf.

Bray, F., *The Rice Economies*. Oxford: Basil Blackwell, 1986.

De Koninck, R., "How Small Peasants Help the Large Ones, the State and Capital", *Bulletin of Concerned Asian Scholars* 15, no. 2 (1983): 32–41.

_____, "On the Geopolitics of Land Colonization: Order and Disorder on the Frontiers of Vietnam and Indonesia", *Moussons* 9, no. 10 (2006): 33–59.

Gourou, P., *La terre et l'homme en Extrême-Orient*. Paris: A. Colin, 1940.

Langner, A., J. Miettinen and F. Siegert, "Land Cover Change 2002–2005 in Borneo and the Role of Fire Derived from MODIS Imagery", *Global Change Biology* 13 (2007): 2329–40.

Parker, C., A. Mitchell, M. Trivedi and N. Mardas, *The Little REDD Book: A Guide to Governmental and Non-governmental Proposals for Reducing Emissions from Deforestation and Degradation*. Oxford: Global Canopy Programme, 2008.

Index